VOLUME THREE

FORMING METHODS

Forming by Hydraulic and Crank Presses

Cast and Molded Dies

1943

reprinted by Lindsay Publications Inc

PROPERTY OF THE

BUREAU of AERONAUTICS

Manufacturing Operations Section

NAVY DEPARTMENT—WASHINGTON, D.C.

UNAUTHORIZED CIRCULATION OF THIS PAMPHLET
BY ANYONE IS PROHIBITED

FORMING METHODS - VOL 3

Forming by Hydraulic and Crank Presses
Cast and Molded Dies
by Bureau of Aeronautics, Navy Department

Reprinted by
Lindsay Publications Inc
Bradley IL 60915

ISBN 1-55918-380-2

2009

1 2 3 4 5

WARNING

Remember that the materials and methods described here are from another era. Workers were less safety conscious then, and some methods may be downright dangerous. Be careful! Use good solid judgement in your work, and think ahead. Lindsay Publications Inc. has not tested these methods and materials and does not endorse them. Our job is merely to pass along to you information from another era. Safety is your responsibility.

Write for a complete catalog of unusual books available from:

Lindsay Publications Inc
PO Box 12
Bradley IL 60915-0012

FORMING BY HYDRAULIC AND CRANK PRESSES

DRAWING OPERATIONS

INTRODUCTION

1. **Drawing Methods.**—The production of aircraft in relatively large numbers makes necessary the application of all types of machines for the fabrication of the parts that go into the planes. The rapid expansion of the aircraft industry has also made it necessary to study those characteristics of aluminum that make it unsatisfactory to apply the same methods of fabrication to that metal as are used for steel. Many of the parts that go into the construction of an aircraft require only a shallow draw or forming, but some require deep drawing or stretching and a careful control of the flow of metal during the forming process. Double-acting dies and double-action presses have in a large measure been the answer to the demand for rapid production of deep-drawn parts.

2. Deep drawing of any piece will cause a compression or shrinkage at right angles to the direction of draw and this compression must be absorbed by either a thickening of the metal or the formation of wrinkles. Aluminum-alloy sheet will stretch readily but has a high resistance to shrinking, and as a result wrinkles will form during drawing unless the dies are especially designed to prevent them. The time element also enters largely into drawing processes since aluminum will rupture if drawn too rapidly, but it can be stretched to about 19 per cent if the drawing process is carried out slowly or in easy stages. Accepted practice sets 8 per cent as the maximum stretch for practical purposes. The dies used for drawing purposes must be designed to allow for any shrinkage that can be foreseen, and the design of the part to be formed should be such that any resulting wrinkles shall be in a portion of the blank that can be trimmed away after forming.

3. The behavior of aluminum during forming under pressure depends upon the nature of the alloy, the design of the die, and the experience of the operator in handling the press. Standard alloys will suffer about the same plastic deformation as pure aluminum but will have a somewhat greater hardness and tenacity. Shapes that are drawn or pressed from an alloy are less likely to rupture in their walls, or at the radius of the punch, than pure aluminum because of these characteristics. An alloy, such as duralumin, or dural, containing from 3 to 4.5 per cent of copper together with a small amount of magnesium, when in an annealed or softened condition will have from two to three times the tensile strength of pure aluminum, but will submit to about the same amount of drawing without rupture. Dural, however, will work-harden and for some conditions it becomes necessary to interrupt the drawing process in order to anneal the metal by heat. The characteristics of different alloys are such that a procedure that will work satisfactorily with one alloy may fail with another. Different shipments of the same alloy may show slightly different characteristics that will only be discovered during drawing or pressing.

4. Controlling Factors in Drawing.—Important factors in the drawing of aluminum alloys are:

(a) The size of the blank compared to the size of the finished piece. A cup may be successfully drawn from *24SO* in one operation if the inside diameter of the finished piece is not less than .6 times the diameter of the blank; if the operation is carried out in three steps using successively smaller dies the ratio may be reduced to a diameter of .4 times the blank.

(b) The pressure with which the edges of the blank are held during drawing. This pressure should be sufficient to permit some slip between the die and the pressure plate, but must be sufficient to cause actual drawing of the metal under the punch. Experience will determine the pressure to use in each case.

(c) The hardness, or temper, of the sheet. The greater the hardness the greater the tendency to rupture and the greater the force required for drawing.

(d) The exact shape of the punch and die. Aluminum is sensitive to the radius of the drawing die and still more so to that of the punch. When the radius of curvature of the punch is too large the tendency will be to perforate the blank rather than to stretch it. If the radius is too small the blank will rupture in the vicinity of the radius because of a refusal to stretch over a too-sharp corner. The radius may be replaced by a 45-degree bevel with suitable radii at the corners, but, as the hardness of the metal increases, the bevel becomes less effective.

(e) The condition of the die face. Cleanliness is highly important because of the relatively soft surface of the sheet. Any dirt on the face of a die or punch will have a tendency to scratch the surface of the sheet and unless this scratch is rubbed, or polished, out it may serve as a point from which a crack may start. If the sheet has been anodized and the scratch penetrates the coating it may serve as a point from which oxidation will start.

(f) The proper lubrication of the sheet. The importance of proper lubrication cannot be over-emphasized. Lubricants will vary from ordinary cylinder oil to compounds in which material such as chalk or paraffin have been mixed with the oil. Special oils are used on the more difficult jobs.

(g) The temperature. Ordinarily, drawing operations may be carried out with the alloy at room temperature but if failure occurs under these conditions, satisfactory results may be sometimes obtained by heating the sheet to a temperature between 100° and 180° C., corresponding to temperatures between 212° and 356° F.

5. Characteristics of the Hydraulic Press.—The load pressure of a hydraulic press can be set at any required maximum so that if for any reason this maximum is reached, the press will automatically release itself before any damage occurs. This characteristic is important in drawing aluminum alloys, which have a tendency to work-harden. As the metal hardens the pressure required to draw it increases. Spoilage may be avoided

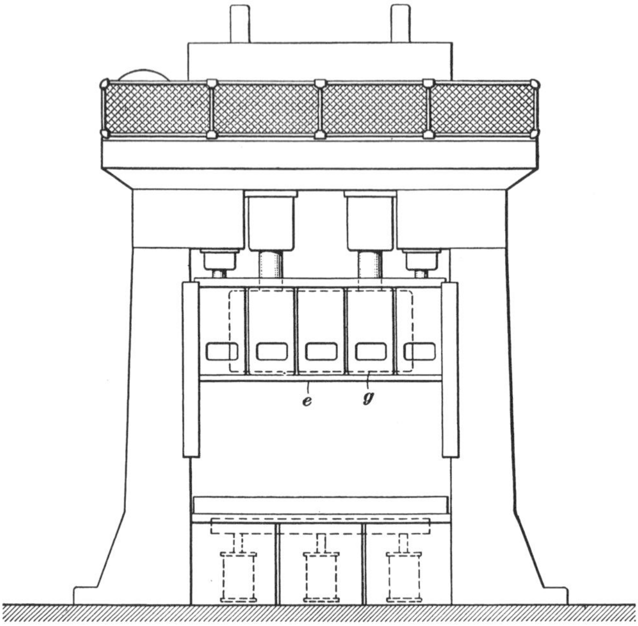

Fig. 1

by setting the release pressure low enough to avoid rupture. Should excessive work-hardening occur the press will automatically release and the partly formed piece can be removed and annealed. Fluid pressure permits of a rapid approach of the punch to the work and of a carefully controlled application of pressure and speed at the time of contact, thus avoiding a blow that might harden the metal. Once contact has been made the speed at which the draw will take place can be controlled, which makes it possible to make a fairly deep draw in one stroke at low operating speed.

6. The low speed of the hydraulic press permits the use of low-cost punches and dies since there is little or no shock to these parts at the time of contact and comparatively little strain during drawing. Also, the stroke is held or dwells at the bottom position which gives the metal time to set and prevents

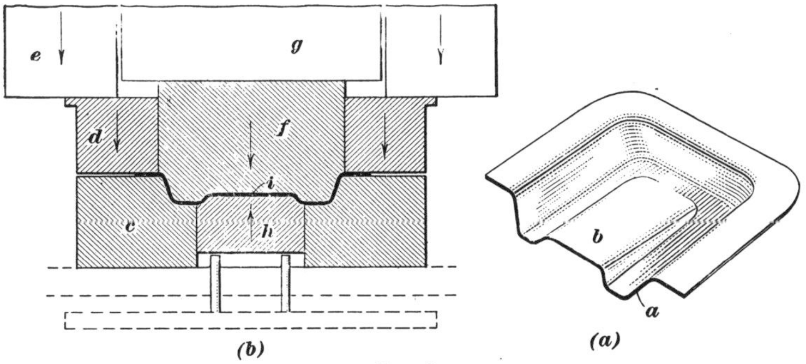

FIG. 2

springback to a considerable extent. The hydraulic press does not have a fixed bottom position so that two sheets of alloy may be drawn at the same time, provided that the work is of such a nature that the final forming is to be done in a separate operation. The action of a hydraulic press may be jerky during the descent of the plunger because of the friction involved, and this condition should be minimized as much as possible to avoid shock that would produce hardening.

7. When the number of pieces to be produced justifies the cost, hardened-steel or mild-steel dies and punches may be used but they must be carefully applied, as they are apt to foul because of the adhesion of particles of the sheet to the steel. Satisfactory results can generally be obtained by using dies and punches made of zinc alloy that can readily be replaced at low cost when they become worn.

TRIPLE-ACTION PRESS

8. Characteristics of Press.—The triple-action press is especially useful in the aviation industry since it can be equipped with Marquette pins in either the ram or in the base. Few of the forms developed from sheet aluminum for aircraft require the third operation as the sheet is generally formed over a solid punch, but the use of the triple-action press makes it possible to attach the die to either the ram or the bed and also to

FIG. 3

operate the pressure, or hold-down, plate from either. A triple-action press is shown in a front and partly diagrammatic view in Fig. 1, and an example of the type of work that can be done on such a press is the pan *a*, Fig. 2 (*a*), in which the bottom *b* is raised to form a trough around the outer edge.

9. A blank that has been cut to the proper size and shape is placed on the die *c*, Fig. 2 (*b*), which is fastened to the bolster plate of the press. The blank is held in place by a pressure, or hold-down, ring *d* attached to the slide *e*, which is actuated from the press head by Marquette pins, and the punch *f* is attached to the ram *g* driven from the main cylinders of the press. The slide *e* and the ram *g* are shown also in Fig. 1. After the blank has been bottomed in the die, a second punch *h*, Fig. 2 (*b*), actuated by a ram in the press bed *i* forces the bottom of the pan into a recess formed in the bottom of the main punch.

10. Corrugated Forms.—The need for a triple-action press is illustrated by the forming of the inner skin of the wing nose shown in Fig. 3, which requires a third action. In this case the need occurs because of the corrugations which must be formed in addition to the contour forming required. The alloy sheet *a* is draped between the rails *b* and is held firmly in place by the hold-down plates *c*, which are loaded from a hold-down ram in the press head to prevent any slippage of the sheet. The amount of drape is accurately calculated or determined by trial, so that the sheet will be stretched to a point at which it will be given a permanent set without over-stressing the metal when the corrugations are formed.

11. The stretch operation brings the sheet to the proper contour and to some extent forms the corrugations *d*, Fig. 3, by stretching the metal over corrugations formed in the punch *e*. The corrugations are given a final and full forming by a die *f*, which is raised against the punch by a ram located in the base of the press, while the punch holds the sheet in the stretched position. The stretch given by both the punch and the die is sufficient to set the sheet.

FIG. 4

12. An example of a triple-action press used for double-action purposes is shown in Fig. 4. In this case the die *a* is fastened to the press bed and the punch *b* is bolted to the head or ram. A hold-down ring *c* is operated from an auxiliary ram in the press head and since a third action is not required, the pins have been removed from the press bed making the ram in the bed inoperative. These pins pass through holes *d* in the press bed when the use of the lower ram is required.

13. When more convenient to do so, the punch is attached to the press bed and the die to the ram. The hold-down plate, or ring, is then actuated from the lower ram and the pins are removed from the head as shown in Fig. 5. In this case the punches *a* are attached to the press bed and the pressure, or hold-down, plates *b* are operated through pins *c* by an auxiliary ram in the press bed. The dies *d* are attached to the ram. The piece to be formed is of such a size that four punches can be fastened on the press bed and four corresponding dies to the head at one time. The position of the punches, and consequently that of the dies, is determined by the location of the pin holes in the bed.

DOUBLE-ACTION PRESS

14. Wrinkling.—During a drawing operation wrinkles will form in that portion of the blank that has not yet passed over the radius of the die, because of the reduction in area of this part of the blank. The tendency to wrinkle can be overcome by the use of a pressure plate in a double-acting press. Since the space between the pressure plate and the die is not sufficient for the formation of wrinkles, the metal must necessarily thicken as the edge of the blank approaches the radius of the drawing die. The pressure exerted by the pressure plate must be carefully regulated. It should be enough to prevent the formation of wrinkles but must not be enough to prevent a uniform drawing of the blank into the die, since this would produce rupture. The drag from the pressure plate if not properly adjusted may cause the flow of the metal to occur in a direction not desired. If wrinkles are allowed to form and are drawn into the die the product will be worthless.

a
b
c
c
c
d
a
b
c
c
c
c
d

15. Puckering.—A wavy formation that has not reached the proportions of a wrinkle is known as puckering. It can be overcome to some extent by increasing pressure-plate loading. The chief factors that enter into a study of wrinkling are: the hardness of the metal, the slipperiness of the lubricant, the size and radius of the die, the thickness of the blank, and the shape of the article to be formed. Aluminum and its alloys pucker badly unless properly controlled. The tendency to pucker increases with a decrease in friction but increases with an increase in radius. Since an increase in the radius of a die reduces the tendency to wrinkle but increases the tendency to pucker, the die design must be a compromise between the two evils. After drawing, the puckers and residual wrinkles may be removed by means of a drop hammer.

16. Die and Punch.—The die used for drawing operations on a double-action press may be merely a draw ring since the blank will form tightly around the punch, which will have the true form of the required part. A gradual development of the draw in steps of about 15 per cent each of the total draw, is sometimes preferable to completing the operation in one step and ironing the walls of the finished shape. Less fouling and breakage will occur when the drawing is done in several operations and, when necessary, the work can be heat-treated in cases where work-hardening occurs. To avoid excessive load on the pressure plate and consequent rupture of the blank, it is customary to set the load so that some wrinkles are allowed to form. The drawn shape can then be "coined" in a drop hammer to the finished shape. During coining the finished shape is accurately sized, the corners are sharpened, and any necessary ridges or indentations are formed. Ironing of the walls will reduce the cup thickness from that of the blank to the clearance between the punch and the die. When possible conical dies can be used to advantage as they reduce the force needed for drawing, and also reduce the tendency to wrinkle and pucker.

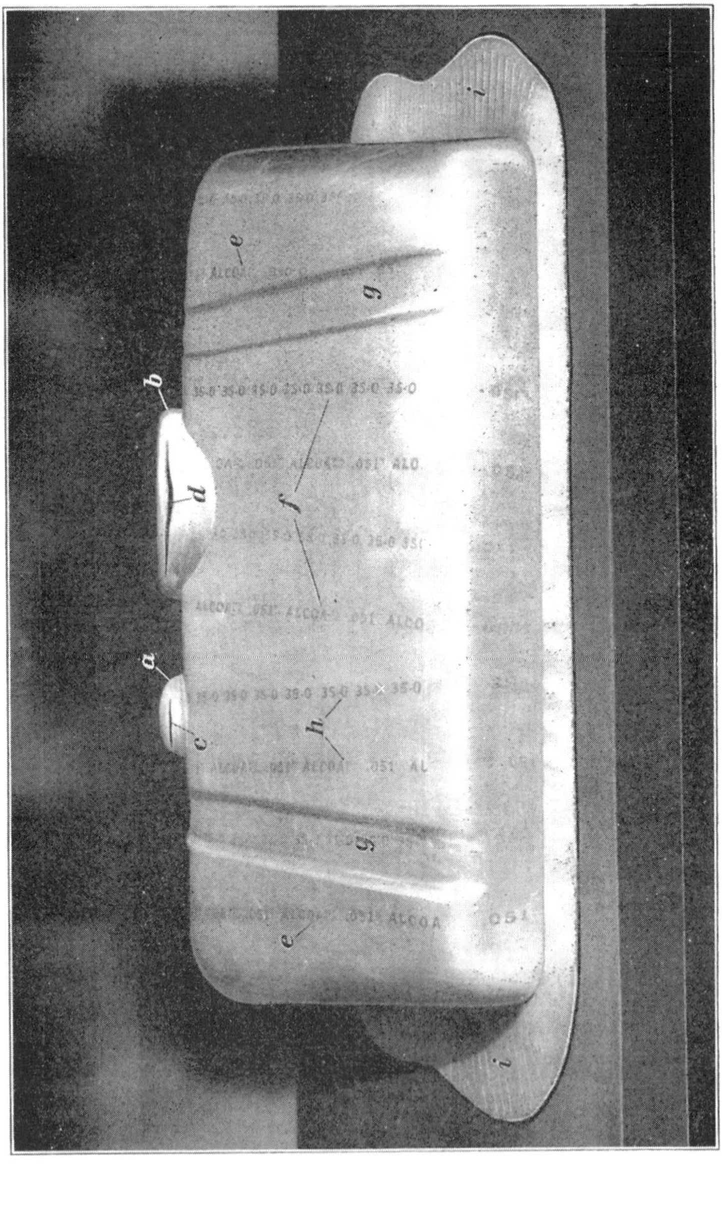

FIG. 6

HYDRAULIC PRESS PRODUCTS

17. Oil Tank.—The oil tank shown in Fig. 6 presented difficulties in deep drawing because of the two outlets *a* and *b* formed on the bottom. One difficulty is shown by the ruptures *c* and *d* indicating that the blank was over-stressed at these points. Trouble of this kind cannot be foreseen and the necessary corrections in the punch and die must be made after test blanks have been drawn. The stretch of the sheet is shown by the position of the printed lines, all of which were originally parallel but which in the finished piece slope in both directions. The end identification lines *e* remain practically parallel to the end of the formed piece, showing that most of the draw was made from that portion of the blank lying outside of the radius of the die.

18. The lines *f*, Fig. 6, slope towards their respective ends of the tank indicating that the formation of the rib *g* and outlet *b* caused the metal to flow in that direction, whereas at the other end the lines *h* slope in the opposite direction. Partly formed wrinkles *i* indicate that the load on the pressure plate was sufficient to prevent serious wrinkles as the blank was drawn toward the radius of the die, but was not sufficient to prevent the necessary slippage of the blank between the pressure plate and the die.

19. Because of the small size of the pressure plate *a*, Fig. 7, it is supported by pins *b* that rest on an air cushion in the base of the press and the punch *c* is attached to a ram located in the press bed. The die *d* is attached to the upper ram. The pins *b* allow a certain amount of depression of the pressure plate under the action of the upper ram, and permit of the adjustment of the pressure between the two parts. The maximum upward movement is limited by lugs on the upper end of the columns *e* that are attached to the press bed. When in operation the lower ram is retracted until the punch lies level with, or slightly below, the level of the pressure plate, the blank is put in place, and the die lowered until the proper pres-

Fig. 7

FIG. 8

sure is developed between the die and the pressure plate. The lower ram then forces the punch into the die, drawing the blank to the desired form.

20. To set up the punch *a*, Fig. 8, and the die *b*, they are placed on the lower ram with the die resting on the punch. The die is attached to the ram by clamps *c*, and the ram is then raised. Prussian blue is spread over the punch and the ram is lowered until contact between the punch and the die is made. The punch is then shifted to avoid contact with the sides of the die. The process is repeated until the punch is properly centered under the die, with the necessary clearance all around to allow for the thickness of the metal, after which the punch is fastened to the lower ram. A test blank is drawn to determine

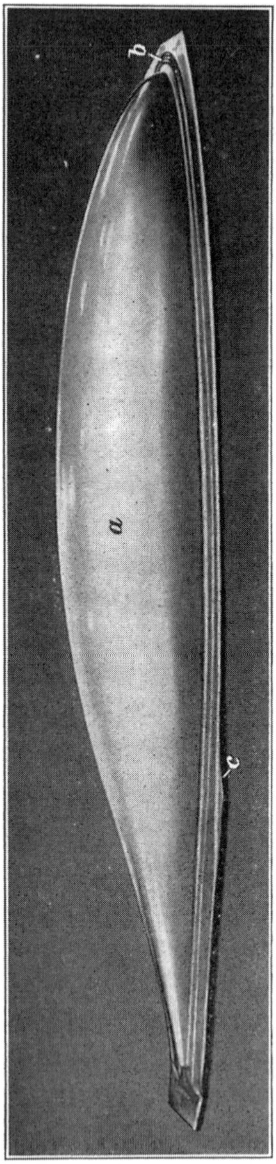

Fig. 9

the accuracy of the setting and the performance of the die. Since the punch is cast in the die, it will be necessary to grind and scrape the required clearance from the die the first time that the set-up is made. If injury occurs to the punch during use or if alterations must be made as indicated by the test drawing, corrections may be made after the punch and die are fastened in place.

21. **Drop Tank.**—The two halves of a drop tank are formed separately and are welded together after the proper bulkheads and baffles have been installed. One half of such a tank, which is formed from *24SO*, is shown in Fig. 9. The blank is cut to the general shape of the half tank *a* and is held in place by a pressure plate during the drawing operation. A bead *b* formed in the edge of the blank helps to control the movement of the blank during the drawing operation without the necessity of excessive load on the pressure plate. The distortion *c* of the edge of the blank indicates the amount of slippage at the area of maximum draw.

22. The punch *a*, Fig. 10, is attached to the ram of the press, and the draw ring *b* is supported at the proper height by blocks *c* resting on the press bed. The pressure ring *d* is attached to pins *e* and is operated by an auxiliary ram in the

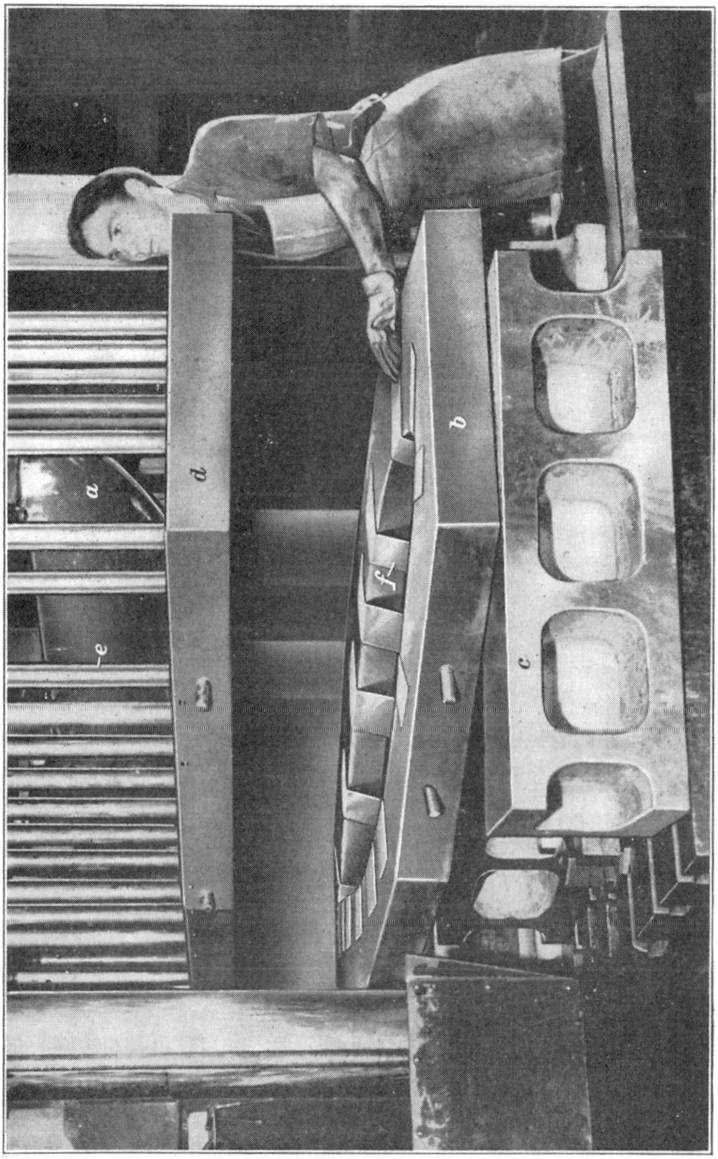

FIG. 10

FIG. 11

head of the press. The opening in the pressure plate is of the same shape as that in the draw ring, but is large enough to permit of the free passage of the punch. When the draw ring has been properly placed on the press bed, the necessary clearance is established by strips of alloy *f* having the same thickness as the metal to be drawn. The punch is dressed until it will fit inside these strips. Once the clearance has been determined, the correct position of the draw ring is established by the use of similar strips when the punch and draw ring are set up for repeat orders.

23. A half tank can also be drawn in a die as shown in Fig. 11. The die *a*, cast of Kirksite, is made with a groove *b* cast in the upper face. A corresponding bead in the pressure plate *c* serves to hold the blank in place as the punch *d* draws the blank to the required form. The die is cast to the exact form, and clearance equivalent to the thickness of the metal is given to the punch. The punch and die are located on the press in the usual way when first set up, and the necessary clearance is ground and scraped from the punch. For subsequent settings, narrow strips of metal having the required thickness are used for locating the die on the press bed.

24. Because of the drawing of the metal with the ring or die, wrinkles *a,* Fig. 12, are liable to form at the ends of the half tank. Since the load on the pressure plate must be uniform throughout, it is practically impossible to establish a load that will permit of the proper slippage at the middle of the half tank, and at the same time be sufficient to prevent wrinkling at the ends. As a result the metal drawn into the sharp bends at the ends of the draw ring or die will tend to wrinkle. These wrinkles can be held to a minimum but, when they do form, they must be removed by hand or machine shrinking. When a failure such as shown at *b* occurs it is generally along the line at which the flange joins the body of the tank, since there will be a sharp corner at this point. Lack of proper lubrication anywhere on the blank will cause an unequal draw on the blank, and set up a strain at some point that will stretch the blank to the point of rupture. It is important that such exaggerated strains be prevented. Since the flanges *c* will be trimmed away the wrinkles here are of no importance.

25. Stretch Forming.—A double-action press may be used in some cases for stretch forming. In such cases no die is used, the ends of the sheet being held firmly between a hold-down plate, or ram, and a rail set on the press bed, as the punch stretches the sheet to the required shape. The punch is formed to the proper shape and during the stretching process the sheet forms tightly to the punch. When forming by this means the

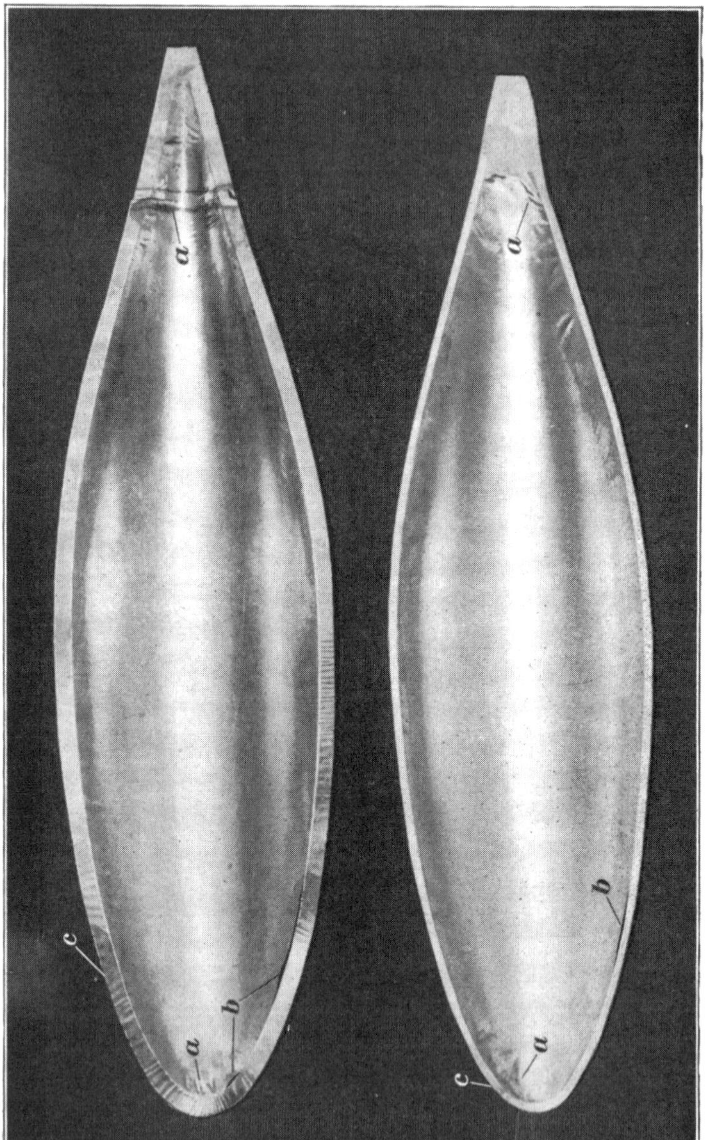

FIG. 12

exact free length of the sheet must be carefully calculated, in order that the proper amount of stretch shall be given to provide a permanent set. At the same time, the metal must not be overstretched to the point at which rupture may occur.

26. Such parts as fuselage and wing skins, which have a double-curvature contour of large radius and shallow depth and which, therefore, require little actual distortion of the flat sheet, can be stretched to a permanent set and with a high degree of accuracy in a double-action hydraulic press. Also such parts as nacelle skins, for which the sheet must be given a definite arc of curvature as well as a double-curvature contour, can be formed by stretching during the formation of the desired curves. Even though a double-curvature contour is required, stretching in one direction only will cause the metal to flow at right angles to the stretch and will therefore produce the required contour. Consequently, the sheet need be gripped only along two opposite sides, leaving the other two sides free to move as the stretch requires.

27. The direction of restraint *a*, Fig. 13, should be parallel to the direction of deepest draw *b* so that the edges gripped will be at right angles to the deepest draw and parallel to the direction of shallow draw *c*. The hold-down plate and rail should have smooth faces as any serrations or other irregularities will cause the sheet to tear. The rails may be set at any angle to conform to the shape of the blank, such as that for a conical section. For flat sections of double-curvature and shallow depth, the

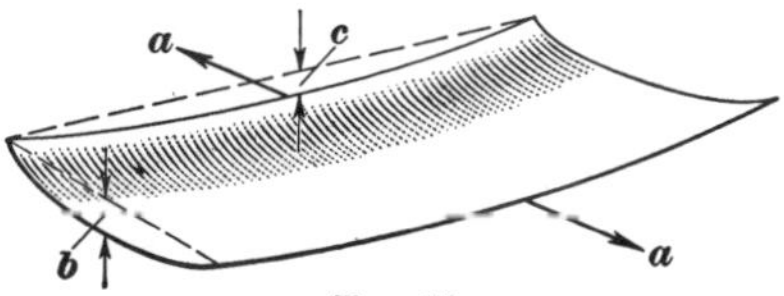

FIG. 13

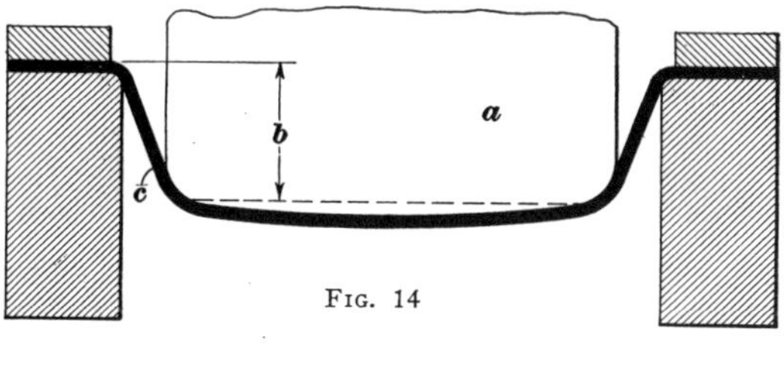

FIG. 14

c
b
d
e
e
a
c
b
d

punch *a*, Fig. 14, may be given additional thickness *b* so that the sag or drape of the sheet *c* can be taken up and the contour formed by the bottom face of the punch.

28. Important Factors in Stretch Forming.—Factors that must be given consideration when stretch forming is contemplated are as follows:

1. Permissable elongation of material.
2. Elongation required to form the part.
3. Shape of the rail over which the material is to be stretched.
4. Radius of rail edge over which the material is to be stretched.
5. Surface condition of the punch.
6. Lubrication.

29. Conditions Governing Stretch Forming.—There are several governing conditions that must be kept in mind if stretch forming is to be done successfully. These conditions are as follows:

1. Maximum elongation for *24SO* and *24ST* is from 8 to 10 per cent with 7 per cent for *24SO*, and 8 per cent for *24ST*, the practical limitations. The elongation of the sheet will be greatest in those sections not in contact with the die, showing that the full ductility of the metal is not available because of the effect of friction on that section that is in contact. The width of the sheet has no effect on the allowable elongation.
2. Generally, an elongation much below the practical limits given will provide permanent set to the metal. The exact per cent required must be found in each case by test.
3. The rails should have straight sides.
4. Radius of rail over which the sheet is to be stretched should be kept as large as possible, and should not be less than 1 inch.

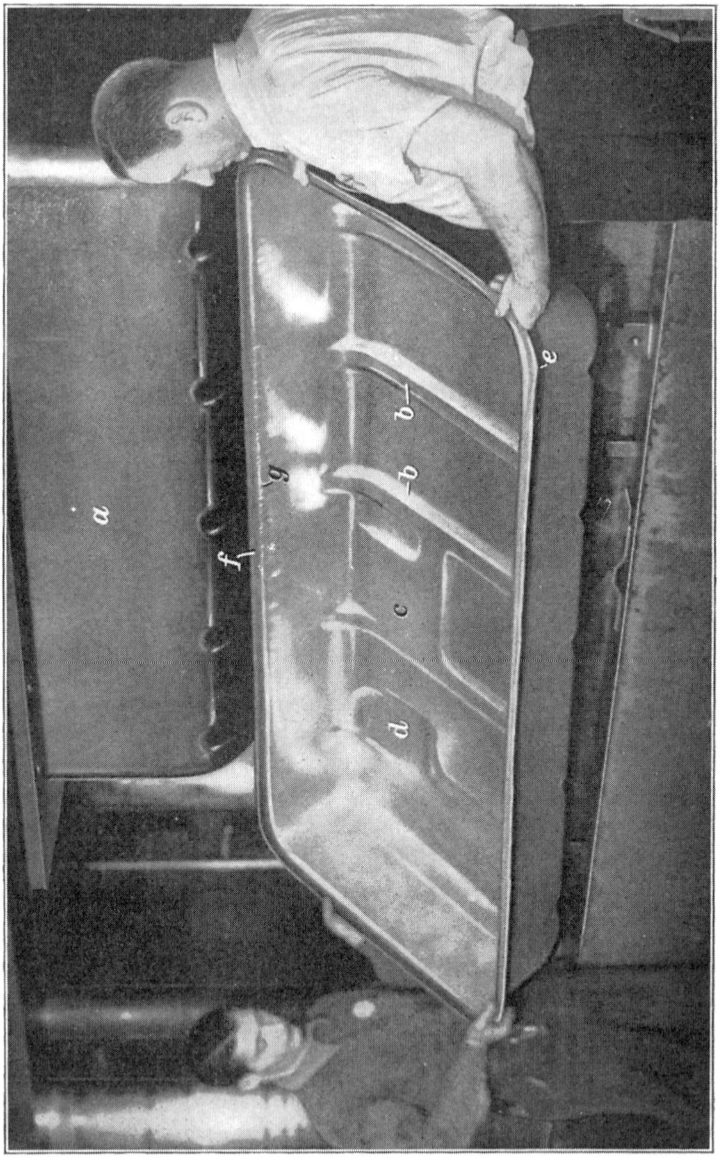
a
b
b
c
d
e
f
g

5. The surface of the punch should be kept highly polished in order to reduce friction to a minimum.
6. Any good drawing oil may be used as a lubricant but it must not break down at any temperature below 150° F., and must provide viscous lubrication at pressures up to 5,000 pounds per square inch.

30. The formation of a nacelle skin lends itself particularly to stretch forming without a die in a double-action press. Pin holes set at a definite center distance apart on the ends of the sheet *a,* Fig. 15, fit over pins set in the rack *b* and the sheet is clamped firmly in place by a hold-down plate *c* under the action of an auxiliary, or hold-down, ram in the press head. The center distance between pin holes may be either calculated or determined by trial, but once it is established, is maintained for all sheets that are to be stretched. This distance is such that the drape of the metal will allow stretching to a permanent set without over-stressing the metal. The correct elevation for the rails *b* is obtained by means of bolsters *d* that rest on the press bed. In order to prevent injury to the sheet while being placed in position, a heavy quilt is frequently placed under the sheet on the press bed. Steel I-beams *e* are bolted to the punch and clamped to the ram head. Originally these nacelle skins were formed of *SO* material, but this method of forming has made possible the use of *ST* material at a relatively high rate of production.

31. Stretch forming on a hydraulic press can be used to advantage in cases where right-hand and left-hand pieces of the same part are to be produced, as has been described in connection with stretch presses. Both the right-hand and left-hand skins can be formed in a single operation by stretching the sheet under a punch. The length of the sheet is accurately determined so that the amount of drape given to the sheet between the hold-down rails is the same in all cases. The punch is so formed that instead of a straight edge of small radius, the apex of the punch is eased to a somewhat elliptical shape allowing the metal to be stretched into the bulge of the skin without

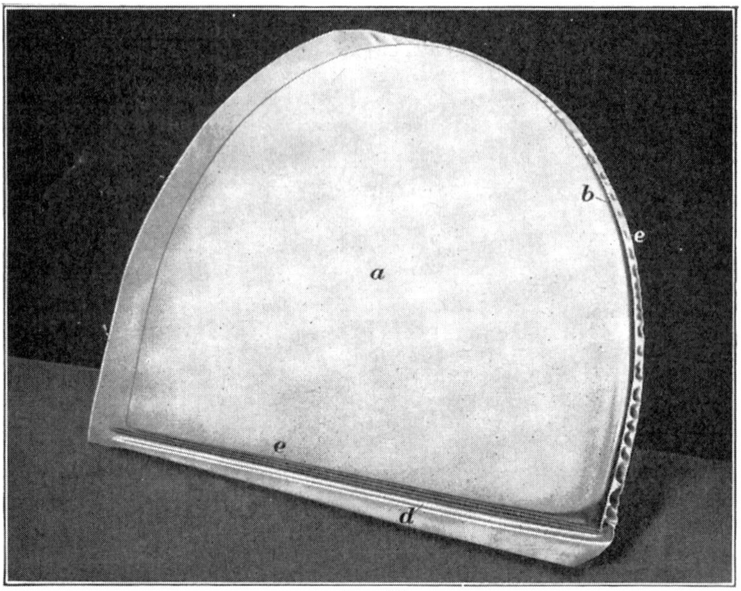

FIG. 17

being over-stressed. The stretching operation is carried only to the point at which permanent set will be given to the metal and springback avoided.

32. **Drawing.**—An example of a large piece formed on a hydraulic press by means of a punch and a die and a hold-down ring is the gas tank, Fig. 16, formed from a single piece deep-drawn to shape. The die, not shown, is attached to the press bed and the hold-down ring is actuated by a ram also located in the base of the press. The punch *a* is attached to the ram. The stiffening ribs *b* as well as the raised section *c* and the depression *d* are formed in the blank during the operation. There will be a tendency for the blank to wrinkle at the corners *e* as the blank is drawn over the radius of the die and wrinkling will occur in the flange *f*, which will be trimmed away. Some wrinkles *g* will be drawn into the sides of the tank and these, together with those formed at the corners, must be removed by hand or by a power shrinker such as a planishing hammer.

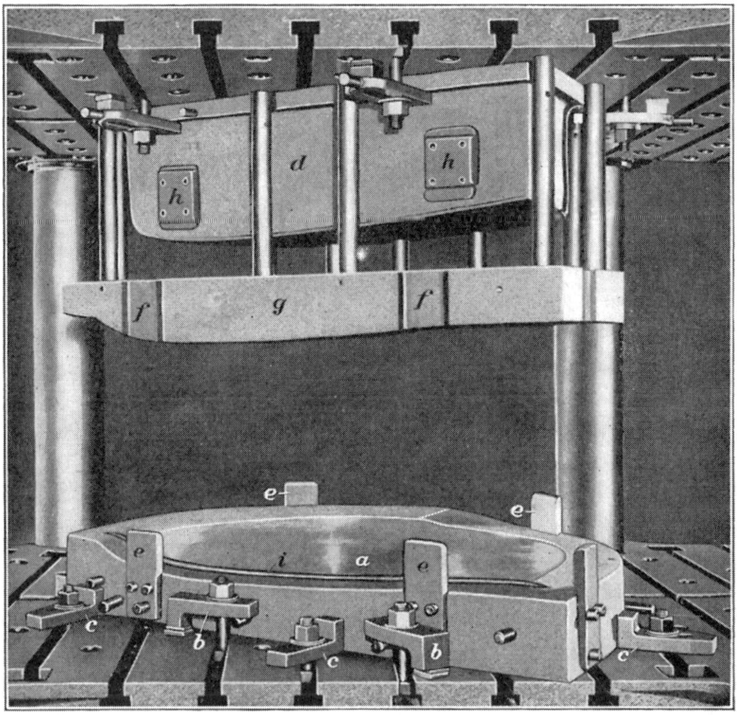

FIG. 18

33. Beading Contours.—The formation of a wing-tip skin *a,* Fig. 17, cannot be completed in a hydraulic press because of an underformed bead *b* that must be developed around a large portion of the curved section. The double-curvature contour on the face of the skin can, however, be formed in a press, leaving the bead for hand finishing in a special jig. During the hand beading, wrinkles are formed in the flange *c* but this flange is later trimmed away at the edge of the bead. A heavy bead *d,* formed in the die, holds the blank in position at one end of the die as the double-curvature contour is formed, so that any flow of metal that may occur must come from the two sides and the other end of the blank.

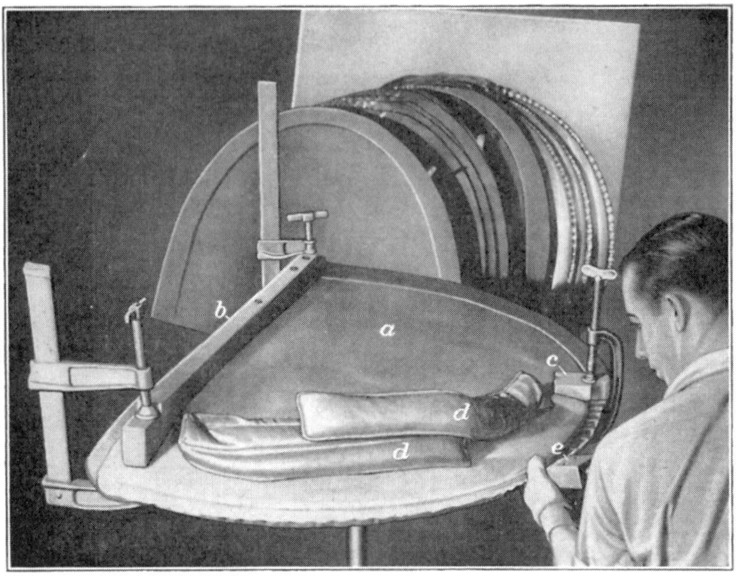

Fig. 19

34. The die *a*, Fig. 18, is placed on the press bed and is held in position by a number of clamps *b*, the free ends of which rest on pins inserted in the die. The face of the die slopes from one side to the other, because of which stops *c* are bolted against the die to prevent a shift in position as pressure is exerted on the die by the punch *d*. Guides *e* attached to the die fit into slots *f* formed in the hold-down plate *g*, and similar guides *h* on the punch fit into corresponding slots on the inside of the plate. These guides insure proper alinement of all three parts during the forming operation.

35. The blank, before forming, covers practically the entire die, lying just free of the guides *e*, Fig. 18, as a considerable excess of metal is required for the development of the hand-formed bead. A heavy bead *i* is welded to the die and fits into a corresponding groove in the hold-down plate. This bead locates one end of the blank leaving the remainder free to slide under the hold-down plate during the forming operation. The exact amount of slippage allowed is controlled by the load on the hold-down plate. The blank will harden during forming, and heat-treatment is necessary before the hand-formed bead can be developed.

36. After forming, the piece is heat-treated and kept in cold storage until the bead is to be hand-formed. The required work must be completed within 20 minutes after removal from cold storage to insure a perfect bead without rupture. The heat-treated skin *a,* Fig. 19, is placed on a special form block and is held in position by a double clamp and a bar *b,* which also hold the straight edge *e,* Fig. 17, in place. Another clamp *c* holds the outer edge of the skin against the form block, and is located at the point at which the bead starts. Shot bags *d* assist in holding the skin flat on the form block so that, as the bead is formed, the skin will be given a slight stretch that will draw it snugly into place and to the exact contour required.

37. Starting at the clamp *c,* Fig. 19, an air-driven or electrically-driven beading hammer *e* is worked several times around the edge of the forming block. During the first passage of the tool a bead is formed only enough to draw the skin into place. Subsequent passages then complete the under-formed bead. This bead is formed only on the leading edge of the skin. The skin for the other half of the wing tip is formed in exactly the same way, and, when put together, the two under-formed beads match and overlap each other perfectly and can be riveted together.

38. A special combined forming block and jig used for a wing tip beaded along the entire edge is shown in Fig. 20. The blank of *24SO* is double-curvature contoured in a hydro-press and placed directly on the forming block *a,* and a grid *b,* which is hinged at *c,* is brought down and clamped in place to hold the formed skin tightly against the form block. The grid is made of Cerro-bend, a bismuth alloy of low-melting point, and is cast and finished to fit the contour of the form block. The use of Cerro-bend makes possible a low-cost grid of a material that is easily cast and machined, and which has sufficient strength

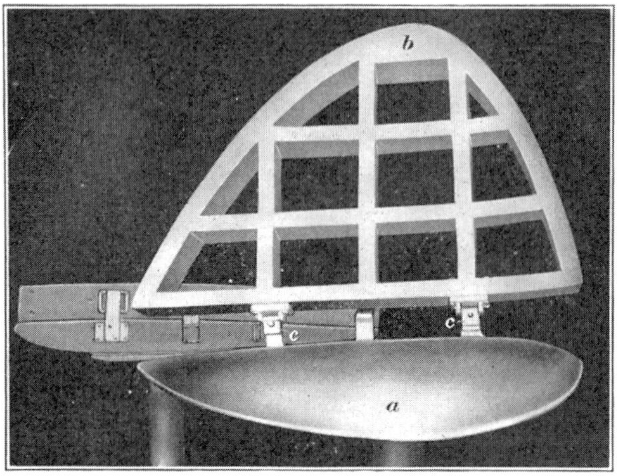

FIG. 20

and rigidity for the work to be done. The Cerro-bend grid is backed up and stiffened by a wooden grid of the same shape and size to give extra support to the metal, since, due to the contour formed, some sections of the metal grid will be thinner than others.

39. A C-clamp *a*, Fig. 21, holds the free end of the grid firmly against the forming block with the contoured skin *b*

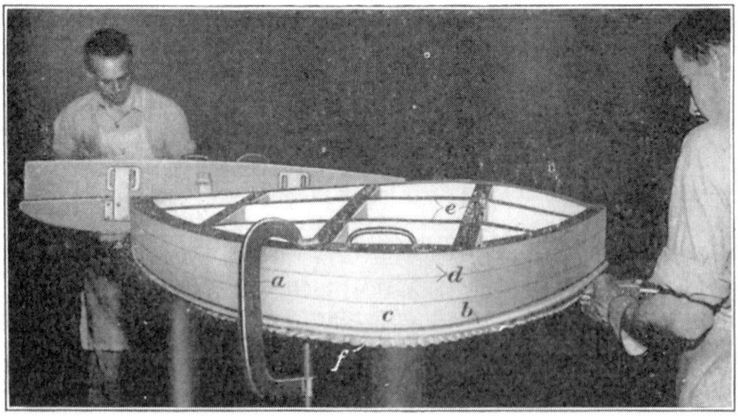

FIG. 21

Fig. 22

between the two. The metal grid *c* is cast to a sufficient depth to retain a thickness of at least ½ inch at the thinnest section. The wooden backing grids *d* are of uniform thickness at the outer edge, the ribs *e* following the contour of the form-block. With the C-clamp in place, an air-driven or electrically-driven beading hammer is worked around the edge of the skins to form the required bead. The grid presses the contoured skin firmly against the forming block, and the bead when formed stiffens the skin so that it will retain the proper contour when removed from the jig. Before the wrinkled edge *f* is trimmed away, the skin is heat treated. The other half of the wing tip is made in the same way. The two halves are assembled, the bead of one fitting snugly in the bead of the other, and the two beads are riveted together.

40. Auxiliary Die.—A double-action press may be used for forming the nose piece of a cowl by means of a special die

FIG. 23

equipped with a movable section, or die, by which the inner section of the blank is curved to form. The punch *a*, Fig. 22, is supported by an air cushion in the base of the press by means of pins *b*, and the die *c* is attached to the ram head. The die *d* has arms, one of which is shown at *g*, that are bolted to the bracket *f*, through which passes the bolt *f′*, which is fastened to the base *e*. A wedge block forces the die *g* toward the die *a* moving the bracket *f* with it against the pressure of the spring *h*. A nut on the outer end of the bolt *f′* limits the outer motion of the bracket *f* and the die *g*.

41. As the press head descends, the die *i*, Fig. 23, forms the blank over the punch *a*, the inside portion *i′* of the die extending far enough down on the punch to force the blank into a cylindrical shape. At the same time a wedge block *j*, attached to the ram head, makes contact with the center portion of the sliding die between the two arms *g*. A further downward movement of the die *i*, against the pressure of the air cushion, forces the sliding die inward, and gives the blank the necessary underforming to force it against the under-curved section of the punch.

42. The die is shown in Fig. 24, in its fully-closed position. The die *i* has forced the punch *a* to its lowest position and the wedge block *j* has forced the sliding die inward or toward the punch. The spring *h* is fully compressed, and will force the sliding die outward releasing the formed piece as the die *i* is raised. Because of the underforming of the nose piece, it cannot be lifted from the punch but must be slid from it endwise.

43. Conical Section. The shape of the piece to be formed governs to a considerable extent the machine on which the job is to be done. A large conical section requiring little contour forming can be shaped to advantage in a double-action hydraulic press, since only sufficient stretch is necessary to set the curve. The sheet or blank from which the piece is to be formed must be bent to the general curve of the punch *a*, Fig. 25, and the flanges formed by hand before being placed in the die, so that the blank will be gripped between the pressure plate *b* and the die *c* before stretching begins. The blank must first be cut to a

Fig. 24

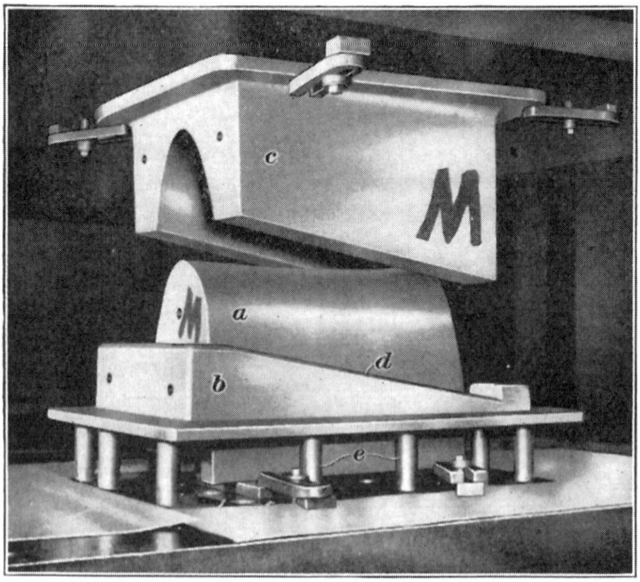

Fig. 25

Fig. 26

Fig. 27

taper and the flanges bent parallel to the tapered edges, or, if this is not done, the flanges must be bent on the sheet to give the necessary taper to fit the sloping faces *d* of the pressure plate. The pressure plate is supported by pins *e* that rest on an air cushion in the base of the press. The flanges should be prebent with a close degree of accuracy along the line of bend, so that the blank will fit to the punch and the same degree of stretch will be given to all pieces formed.

44. Gillotine Die.—The leading edge *a*, Fig. 26, of the contoured cowl sheet *b* must be bent over a too-short radius to be formed when the sheet is given its double-curvature contour, and consequently the forming of this bend becomes a separate operation. The rear end of the contoured sheet rests on the stand *c*, and the end to be bent rests on the die *d* on which it is held in position by pins. A pressure plate *e* actuated by an auxiliary ram in the press head locks the sheet in place, and the punch *f*, attached to the main ram, forms the bend. The face *g* of the die slopes downward and it is necessary to support the punch *f* by means of a backing plate *h*, which forms part of the die, in order to prevent endwise movement. Both the die and the punch are made of a zinc alloy, generally Kirksite, which is sufficiently strong to do the required work. The safety blocks *i* are placed between the press bed and the upper platen during the installation of the die, and at times during which repair work must be done on the die.

45. Multiple Set-Up.—When relatively small dies are used, it is frequently of advantage to set up more than one die on a press bed. In order to do this successfully, care must be taken to adjust all dies and punches to a height that will give full forming to each part for the same length of stroke. When the number of parts justifies the expense of extra punches and dies, two or more sets may be made, depending on the size of the die relative to the size of the press bed. By using a zinc alloy of low melting point such as Kirksite for the die and either Kirksite or lead for the punch, such sets can be made economically.

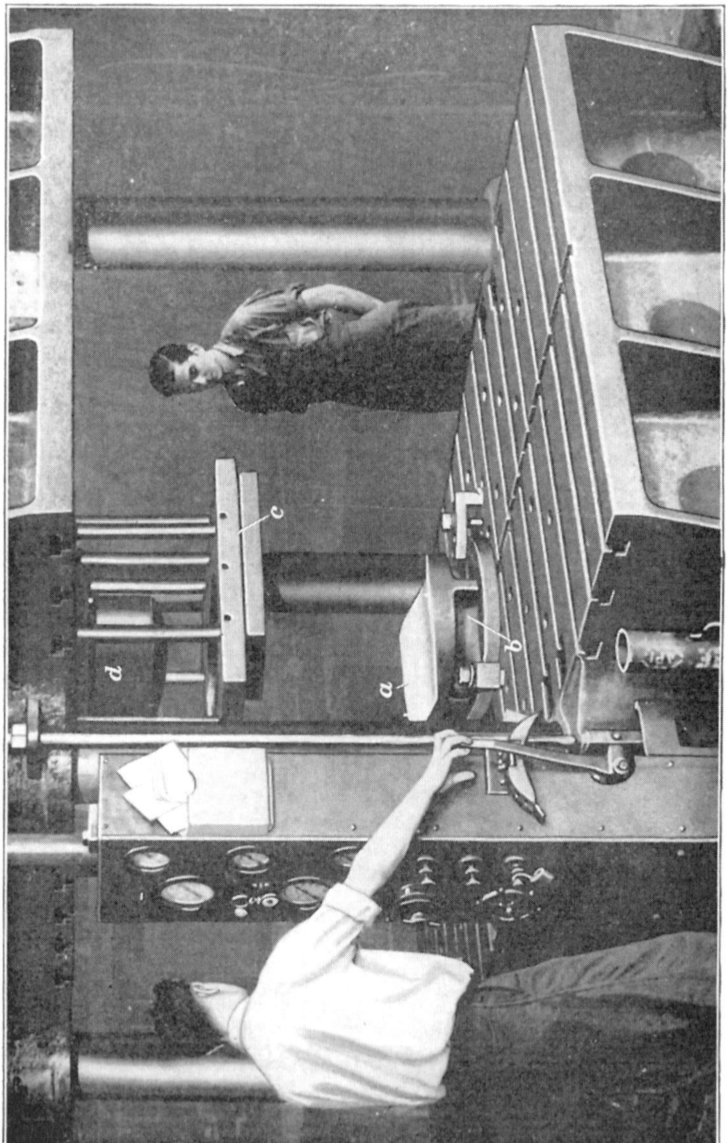

Fig. 28

46. Four sets of dies *a* and punches *b*, Fig. 27, for forming relatively deep tanks are grouped together. The punches are clamped to the press bed and the dies are suspended from the ram. The blanks are placed on hold-down plates *c*, operated by an auxiliary ram in the base of the press, and are held in place by the load exerted by the die. The proper pressure to allow the necessary slippage of the blank is maintained by regulating the air cushion. A triple-action press may be used for this operation, in which case the hold-down plate would be actuated by a ram in the press bed. Hollow metal safety columns *d* are placed between the upper platen and the press bed whenever the ram is to be held in its upper position.

47. Oversize Press.—Each machine should be used to its full capacity as far as possible, but it is frequently necessary to use a large machine for a relatively small job. It is better to do this than to have a machine standing idle and work held up because a machine of the required capacity is not available. While a large press may be used for small work, care should be taken not to overload a small press by forcing it to do work beyond that for which it was designed. An example of a large press doing a relatively small job is shown in Fig. 28. The blank *a* is bent by hand to conform to the general curve of the die *b* and is held in place by the pressure plate *c*, which is actuated by a ram in the press bed. The punch *d* forces the blank into the die, drawing the metal to the required shape.

SINGLE-ACTION PRESS

48. Collector Ring.—An example of an unusually large forming operation is the collector ring, Fig. 29, in which a stack of the formed rings is shown. Holes *a* are prepunched in the blank and serve to locate it by fitting over pins set in the rim of the die. A holddown plate or ring also fits over these pins and forms a crimp *b* in the outer edge of the blank, so that it is held firmly in place during the forming operation. Stiffening ribs *c*, and such clearance offsets *d* as are required, are pressed into the blank during forming. The blank is placed on the die

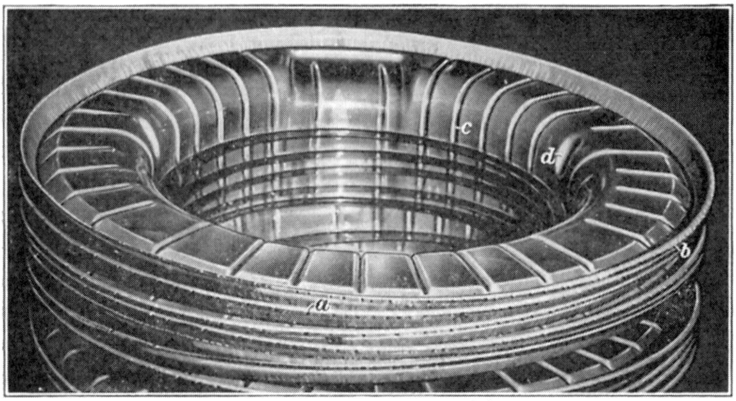

FIG. 29

a, Fig. 30, and the holddown ring *b* is fitted in place. As the punch *c* descends, a spring-loaded pressure plate *d* first contacts the holddown plate, forcing it down over the pins *e,* and forms a crimp in the outer edge of the blank. The punch then forces the inner portion of the blank downward into the die and presses it firmly against the inner wall. Because this is a stretching operation, and because the stiffening ribs absorb any shrinkage that may occur as the blank is drawn over the radius of the die, no wrinkles are formed.

49. The punch *a,* Fig. 31, and the spring-loaded pressure plate *b* are suspended from the ram of the press. The pressure of the springs *c* is sufficient to form the crimp in the outer edge of the blank and to hold it firmly in place. The punch and the pressure plate are bolted together to form a single unit. The ribs *d* and the forms *e* for making depressions in the finished part may be either cast on the punch, or cast separately and brazed onto the die. Holes *f* are provided in the punch to reduce its weight, and the center of the punch is cored out for the same purpose. A circular opening is cut in the center of the blank to fit the leading edge *g* of the punch, and the only trim necessary is the removal of the crimped edge. The same part can be formed in a double-action press, in which case the spring-

Fig. 30

FIG. 31

FIG. 32

loaded pressure plate and the crimping ring would be replaced by a ram actuated holddown plate.

50. Auxiliary Pressure Ring.—A single-action press may be made to function as a double-action press by the addition of auxiliary oil cylinders *a*, Fig. 32, placed around the periphery of the die, or punch, as the case may be. Each cylinder has a close-fitting outer shell *b* and is filled with oil through a connection *c* from a single source of supply. Equal pressure is exerted in all of the cylinders by a single pump at the source so that all of the outer shells exert the same upward force. A pressure ring *d* covers the entire group of cylinders, and both equalizes the pressure and presents a continuous surface against which the blank is held. The cylinders are not attached to the press bed and may be moved to any desired position. The leakage of oil is small if the cylinders are properly fitted. The punch *e* is used for forming an engine cowling.

51. When in the raised position the cylinders *b*, Fig. 32, hold the pressure ring *d* just below the top of the punch *e*, so that when in place the blank rests on the die rather than on the

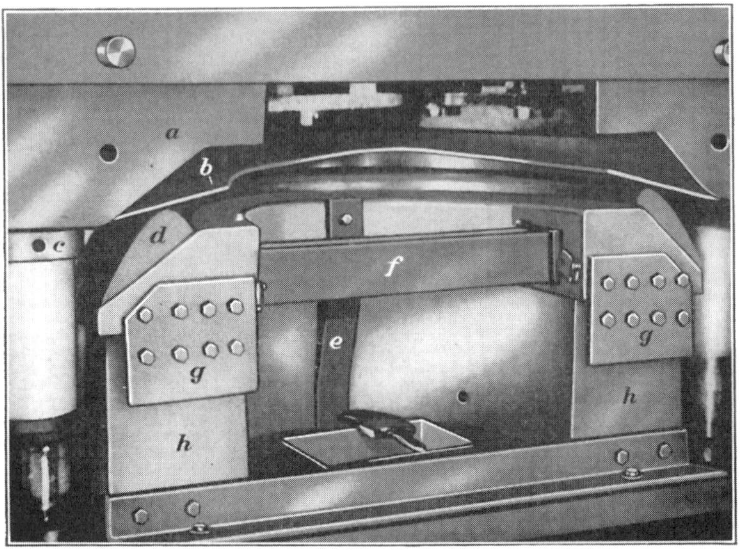

FIG. 33

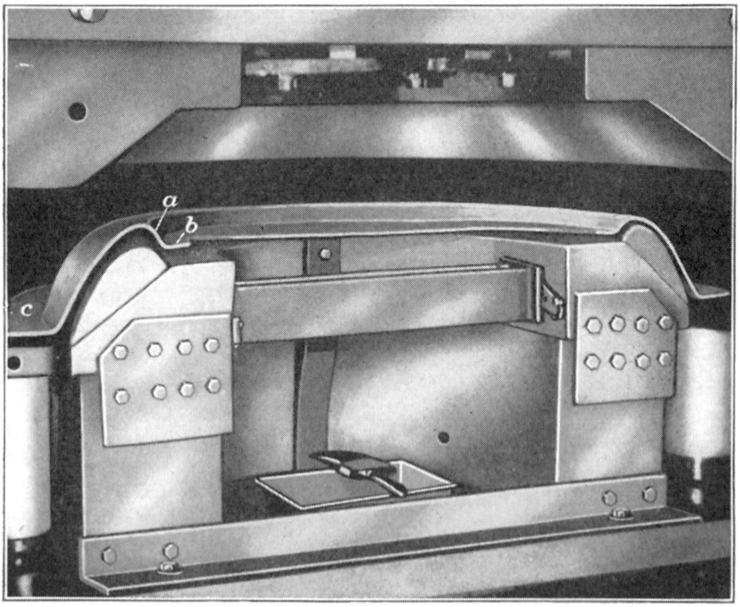

FIG. 34

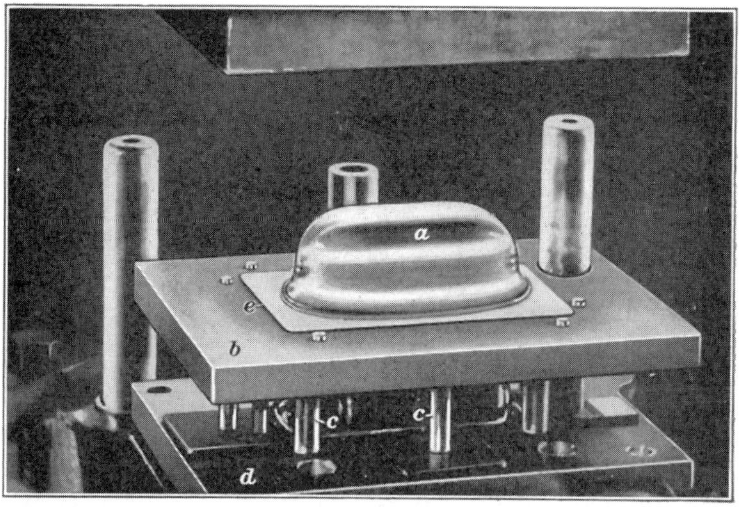

FIG. 35

ring. The number of cylinders used and their position is largely determined by the nature of the piece to be formed. They may be evenly distributed as in the case of the cowl ring or if the nature of the work requires it, they may be grouped more closely at one point than at another. The control of the oil-pump pressure determines the amount of slippage allowed the blank during the forming process if any.

52. As the die *a*, Fig. 33, is brought down it presses first on the outer edge of the blank *b* forcing it down against the pressure ring *c*, thereby distorting the blank to some extent. The slope of the blank outside of the punch *d* helps to prevent excessive slippage as the blank is bent over the nose and down against the inner face of the punch, and gives some slack to the blank. The punch, which is made of Kirksite, is reinforced by a strap *e* and a brace *f* that prevent distortion under the pressure of the die. The strap and the plates *g* hold the punch to the base. If and when distortion or wear of the punch occurs, the nose may be removed from the base *h* and either repaired or replaced.

FIG. 36

FIG. 37

53. The finished piece *a*, Fig. 34, forms one half of the cowl nose and will be riveted to the other half to form the complete nose. The skirt of the cowl, also made in sections, will then be riveted to the nose piece and the sections riveted together. To complete the nose section the flanges *b* and *c* will be trimmed away along a line drawn against a templet that fits over the piece, or the nose piece may be placed in a special jig and the excess material removed by a hand router.

54. Deep Drawing.—A deep-drawn piece *a*, Fig. 35, can be formed on a single-action press by using a spring-loaded ram in the base of the press and a pressure plate *b* supported on the ram by pins *c* that pass through a base plate, or die shoe, *d*, which is bolted to the lower platen of the press. A blank *a*, Fig. 36, is first stamped out to the shape shown. In order to allow for the drawing of the metal, the sides of the blank are curved, providing enough extra metal at the points of greatest draw for the drag of the metal into the die. The die *b* clamps the blank firmly to the pressure plate *c* and draws the metal downward over a die that is attached to the die shoe. The pressure on the plate *c* is sufficient to prevent the formation of wrinkles but allows some slippage of the blank. If the blank is properly designed the flange *e*, Fig. 35, around the finished piece will have straight sides.

55. A deep-drawing operation, that is rendered more difficult by the square corner formed at the bottom of the draw, is shown in Fig. 37. The punch *a* is mounted on the bed of the fixture *b*, or on the bottom platen of the press. A pressure pad *c* is mounted on the clamping pins *d* which pass through the bottom platen and bear against a pneumatic cushion plate beneath the platen. The blank *e* to be formed is placed on the pressure pad *c*, and as the moving platen descends, the female die *f* clamps the sheet-metal blank against the pressure pad *c*. Further downward movement forces the blank over the punch *a* on the bottom platen, and fabricates the part, an example of which is shown at *g*. The pressure of the pressure plate *c* may be regulated by varying the pressure of the air in the cushion plate.

56. In one plant considerable trouble was encountered in the fabrication of the shape shown at *g*, Fig. 37, because of breakage along the upper end of the drawn section at the point *h*. It was found upon examination that the drawing tool used did not possess the finish usually given to equipment of this kind. This fact was borne out by the presence of heavy drag marks at certain points on the flange. The trouble was overcome by giving a high polish to the tool surface and allowing a slightly more liberal radius on the die. Because of the shrinkage of the blank as it was drawn into the die, wrinkles *i* were formed in the flange, which is later trimmed away.

CRANK PRESS

57. Characteristics of Crank Press.—The chief characteristic of the crank press that effects its use in forming aluminum, or aluminum-alloy, sheet, is the constantly changing velocity of the punch which moves according to the law of simple harmonic motion. As a consequence of this action, the metal cannot be coaxed into its new shape as it can by gentle pressure, but must be formed quickly or it will rupture to relieve the strain of deformation. The crank-press action is not affected by the characteristics of the metal being drawn, but the press treats all metals alike and produces a variation in the flow of the metal because of the changing punch speed. Since aluminum and its alloys should be drawn gradually and evenly in order to get the best results, this feature limits the use of the crank press in the forming of aircraft parts. The variation in flow due to varying punch speed prevents taking full advantage of the capacity of aluminum to deform within the plastic range.

58. The action of the press is usually transmitted to the metal at a time when the punch is moving at its greatest speed, so that the effect is that of a blow rather than one of pressure. Dies for forming alumnium on a crank press must be designed with a view to allow for sudden deformation, and, after trial pieces have been struck, some changes or corrections may be required. The final die, therefore, may be one based on experience rather than on the original design.

59. The same characteristics of the crank press that restrict its use in forming, add to its usefulness for punching and piercing sheet stock. There, the suddenness of the blow causes a quick rupture of the metal along the shearing edge of the die, and prevents the drawing of the metal over the edge. Normally the punch will cut about one-third of the way through the metal, the remainder being ruptured by pure shear.

60. Size and Capacity of Crank Press.—The tonnage of a cutting die, or the maximum pressure required to cause it to cut a blank, is found by multiplying the shearing strength of the material to be cut by the area sheared, and dividing the product by 2,000. In other words,

$$T = \frac{Stl}{2{,}000}$$

in which T = tonnage of die;

S = shearing strength of material, in pounds per square inch;

t = thickness of material, in inches;

l = length of contour of the blank, in inches.

If the blank is round, and of a diameter d, the length l is $3.1416\,d$. The shearing strength S of the material may be found by referring to a table of values of the shearing strength of materials. The approximate values of S for some of the common metals are as follows:

Material	Shearing Strength Lbs. Per Sq. In.
Aluminum	12,000
Copper	30,000
Brass	35,000
Low-carbon sheet steel	50,000
High-carbon sheet steel	75,000

The foregoing formula applies only to flat punches and dies that do not have shear. The effect of shear is to reduce the length l in contact with the cutting edge at any single instant and thus to reduce the pressure required.

61. Tonnage of Drawing Dies.—In calculating the tonnage required for a drawing die for drawing a cylindrical sheet, the formula of Art. 60 is used to determine the pressure required to force out the bottom of the shell. In other words, the pressure required to draw a shell is approximately the same as that required to blank out a disk having a diameter equal to that of the shell. If the pressure were any greater than this, the bottom would be pushed out before the shell was drawn. Actually, the pressure must be slightly less than that required to shear out the bottom, in order that the material shall not be thinned out to such an extent that it is on the verge of cracking or tearing.

It is assumed that there is sufficient clearance between the punch and the draw ring to allow the material to flow into the shape of the shell without excessive ironing or burnishing. If the clearance is too small, a large part of the pressure is required to iron out the material to the desired wall thickness and greater pressure is required to draw the shell to the proper depth. The thickening of the wall near the top of the shell, as the greatest area lying outside the radius is compressed and drawn into the die, in this case is what determines the depth to which a shell may be drawn in one operation, for a given material. The polish of the draw ring, the lubricant used, and the speed of the drawing punch also affect the tonnage required and the maximum depth of draw.

62. Estimating Capacity of Press.—After the tonnage of the die has been determined, the capacity of the punch press necessary to operate the die may be selected. It is the practice in many shops to have the manufacturer's rated capacity marked on each press, as a precaution against overloading the press; but the capacity of any particular press may be found by referring to the manufacturer's catalog. In the absence of the foregoing information, the approximate capacity, in tons, may be estimated by multiplying the square of the diameter of the crankshaft of the press by a constant the value of which depends on the construction of the press and the length of the stroke.

In the case of single-crank presses of small and medium sizes, in which the stroke does not exceed the diameter of the crankshaft, the value of the constant would be 3. Thus, if a press of this class had a crankshaft 4 inches in diameter, its approximate tonnage capacity, according to the rule just given, would be $4^2 \times 3 = 4 \times 4 \times 3 = 48$ tons. A large press with double cranks and a stroke from 1 to 2 times the diameter of the shaft would have a constant of about 5. Thus, if such a press had an 8-inch crankshaft, its approximate capacity would be $8^2 \times 5 = 8 \times 8 \times 5 =$ 320 tons. It must be remembered that this method gives only approximate results, and a liberal factor of safety must be allowed. The greater the ratio of stroke to crankshaft diameter for a given type of press, the smaller should be the value of the constant.

63. Lubrication of Dies.—The amount of wear on the dies determines the kind of lubricant to be used. Light blanking and some forming operations that do not produce much friction in the die will require no lubricant. For deep-drawing operations, lard oil is quite frequently used. There is considerable friction between the material and the draw ring in a drawing operation, and good lubrication is necessary to prevent the draw ring from scoring or from picking up material and producing scratches in the drawn shell.

The lubricant may be swabbed on with waste wrapped around the end of a stick, or painted on with a brush. Sometimes the material to be drawn is run between felt rollers soaked in the lubricant, just before reaching the die. In drawing operations, the blanks or shells may be dipped in a pan of lard oil before being placed in the die.

There are numerous lubricants on the market, consisting largely of combinations of mineral oils, vegetable oils, and animal oils. Some of these produce exceptionally good results in reducing wear on dies. In some cases paraffin is mixed with the oil.

64. Effect of Thickness of Material.—Dies made for shaping operations must allow for the thickness of the metal between the upper and lower parts of the die. This is illustrated

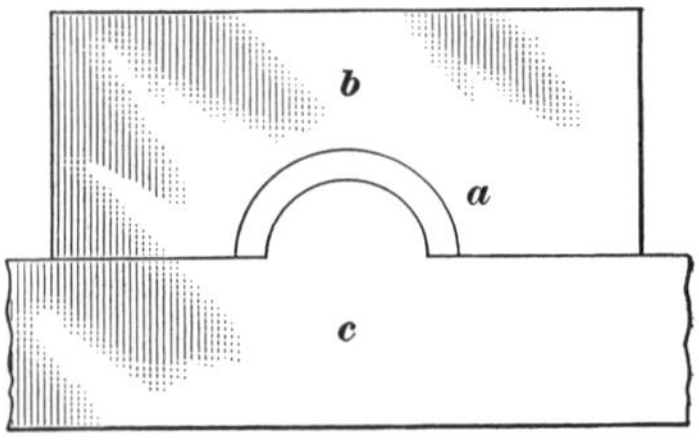

FIG. 38

in Fig. 38, where a thick piece of material *a* is shown between the bending surface of the die *b* and the lower section *c*. The curve of the upper die must have a radius equal to that of the lower die increased by the thickness of the metal. It must also be kept in mind that any material will bend more easily around a curve than over a sharp corner, and that there is less liability to rupture over a curve. All corners of a bending surface must be rounded off to a smooth curve. A thin ductile material, *SO*, can be bent over a smaller radius than harder or less ductile material, *ST*.

65. Bending Dies.—A right-angle bend of the type shown in Fig. 39, view (*a*), may be made by the plain bending die illustrated in view (*b*). The piece of stock *a* that is to be bent is sheared off to the proper length in a preceding operation, and is placed in the nest formed by the two locating stops *b* screwed

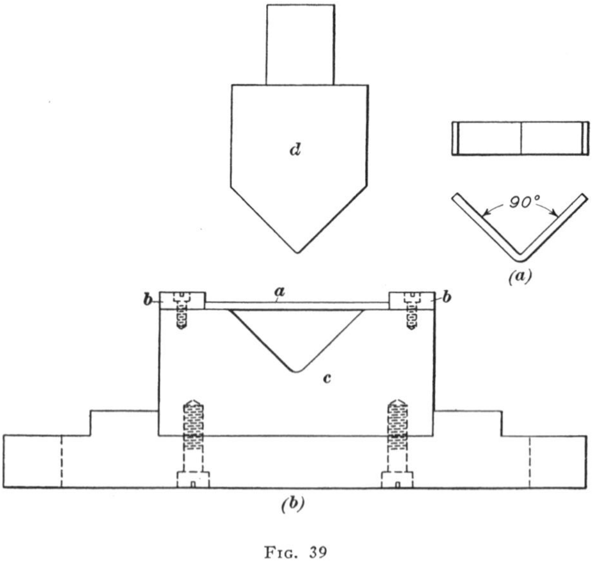

FIG. 39

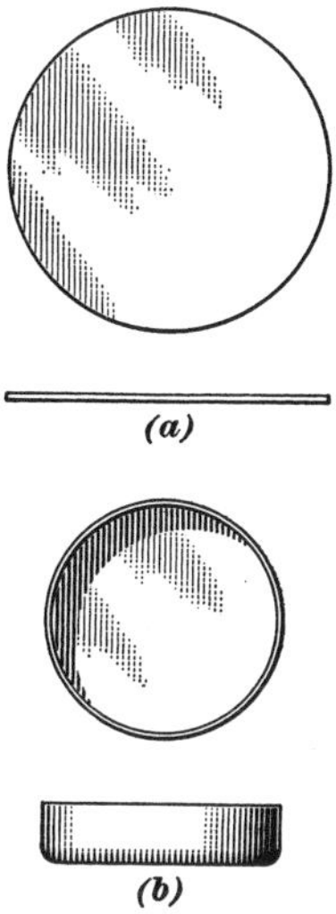

FIG. 40

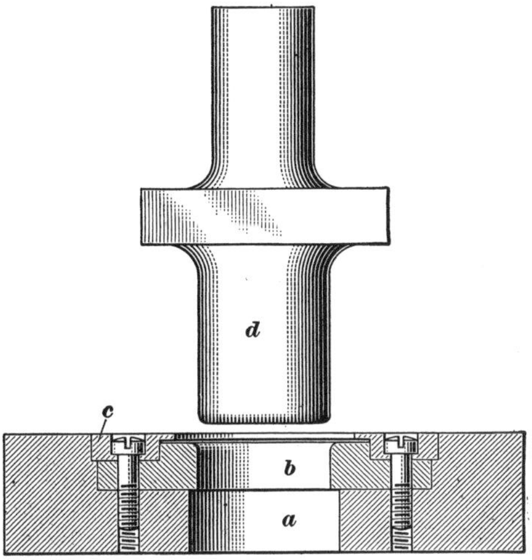

FIG. 41

to the top of the die block *c*. The punch *d* strikes the piece, forces it down into the die, and bends it to the desired shape. In making a die of this type for bending thin, soft material, the angle of the punch and the die may be the same as that desired in the finished piece. For heavy or hard material it is necessary to make the angle slightly less, to allow for a slight springback of the material after bending. The proper angle to which to make the die and the punch in that case depends on the kind, thickness, and temper of the material to be bent and must be determined by trial. The angle for a bending die for phosphor bronze or spring steel, for instance, must be considerably less than that required to make the same bend in soft aluminum. A slight radius must be left in the corner of the bend to prevent fracture of the material. The proper radius depends on the thickness and temper of the material to be formed, as well as on the sharpness of the angle. In general, however, a radius equal to the thickness of the material, or slightly more for hard materials, will prove sufficient.

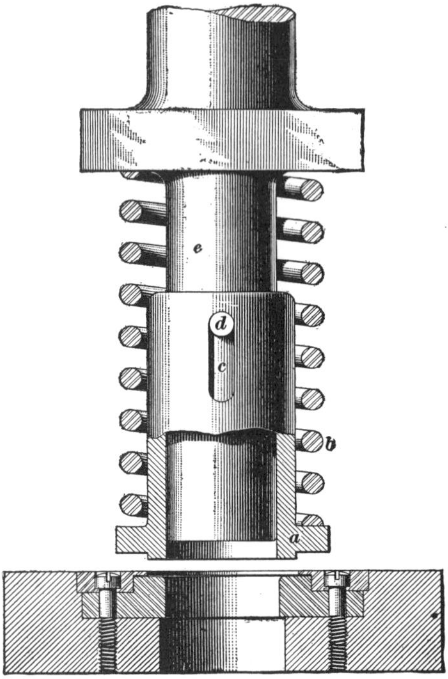

FIG. 42

66. Forming Dies.—A simple example of forming that is applicable to aluminum in crank-press dies is shown in Fig. 40, in which a flat circular blank, view (*a*), is drawn into a shallow cup, as shown in view (*b*). The die used for this operation consists of a die shoe *a*, Fig. 41, bolted to the press bed, a die *b* in which the cup is drawn to size and shape, a gage ring *c*, that holds the die in place, and a punch *d*, that is attached to the press ram. The ring *c* is bored out centrally to the size of the blank, and serves to center the blank in its correct position. The correct position of the ring is assured by boring out the under side to fit a collar formed on the die.

67. The diameter of the punch *d*, Fig. 41, is made equal to the inside diameter of the cup to be formed, which, in turn, equals the diameter of the die less twice the thickness of the metal. The blank is inserted in the gage ring of the die, so that the punch, as it descends, bends up an outer zone of the blank, and in passing through the die straightens out any wrinkles that form. The punch with the work on its lower end, passes completely through the die. The upturned edge of the cup springs slightly away from the punch, from which it is removed by contact with the lower face of the die as the punch is withdrawn. The upper inner edge of the die must be rounded off to a radius somewhat greater than the minimum radius allowable for the metal to be formed.

68. A cup having any considerable depth will have such a wide border around the edge of the die that wrinkles are sure to form as this border metal is drawn into the die, unless some means is provided to control the flow of the aluminum during the shrinking process. To avoid wrinkling, the punch may be provided with a spring-loaded holddown plate, or blank holder, *a*, Fig. 42, that exerts enough pressure on the blank to prevent the formation of wrinkles, or to iron out such wrinkles as may have a tendency to form. The spring *b* holds the blank holder in its lowest position so that the upper end of the slot *c* is pressed against a pin *d* set into the punch *e*, thus allowing a free upward movement of the blank holder after contact has been made with the blank. The pressure of the spring on the blank holder should be sufficient to prevent the formation of wrinkles, but at the same time must allow some slippage of the blank as it is drawn into the die.

69. Size of Blank for Formed Parts.—A general rule that may be considered to apply to all formed and drawn stampings is that the blank should have an area equivalent to the area of the formed or drawn part. If a finished piece is at hand, the weight can be checked with the weight of the blank, and if the measurements are accurately made, the two weights should check within 5 per cent, which is the commercial tolerance for the weight of sheet metal. In determining the equivalent area of the blank, allowance should be made for bends, and for trimming the finished piece when necessary. It is customary to add an allowance of one-third the thickness of the material for each bend in a formed part. For example, if the piece has a single bend, the length of the blank is $l=a+b+\frac{1}{3}\ t$. If the piece has two bends, then $l=a+b+c+2\times\frac{1}{3}\ t$. If it has four bends, then $l=a+b+c+d+e+4\times\frac{1}{3}\ t$, in which the letters represent the length of the sides produced. For any larger number n of bends, the allowance to be added is $n\times\frac{1}{3}\ t$.

70. Procedure for Developing Blank.—When several operations are required to produce a formed part or when the blank from which the part is to be formed must be cut approximately to shape and size before forming, it is necessary to develop the blank by working back from the final forming operation. The proper procedure is, first of all, to make the final forming die complete, according to the dimensions on the drawing. Two similar blanks should then be made by hand, as near to the actual shape required as can be approximated by the methods previously described. One of these blanks should be tried in the final forming die, and the resulting formed part will show where changes in the shape of the blank are necessary. These changes should be incorporated in two more similar blanks, the first unformed blank being used as a templet in making the changes. Then one of the corrected blanks should be formed in the die, and if changes are required, a third set of blanks should be made. This cut-and-try procedure should be continued until a blank of the proper shape is obtained. The blanking die can then be made, the unformed blank from the final tryout being used as a layout templet.

71. Size of Blanks for Drawn Parts.—The only satisfactory way of determining the size of flat blank required to make a drawn part is by trial. However, the approximate diameter may be found by calculation. Table I shows sections of drawn shells of various forms, and formulas for finding the approximate diameters of circular blanks for such shells. It is assumed that the area of blank required is equal to the area of surface of the drawn shell. Therefore, the surface area of the shell is calculated, and the diameter of the blank is then taken as the diam-

TABLE I

DIAMETERS OF BLANKS FOR DRAWN SHELLS

Form of Shell	Diameter of Blank
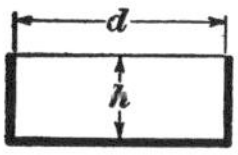	$\sqrt{d^2+4dh}$
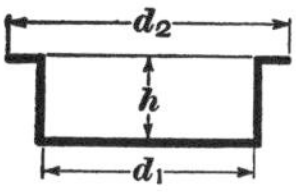	$\sqrt{d_2^2+4d_1h}$
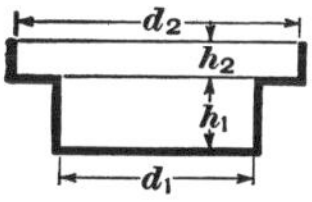	$\sqrt{d_2^2+4(d_1h_1+d_2h_2)}$
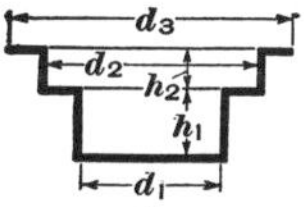	$\sqrt{d_3^2+4(d_1h_1+d_2h_2)}$
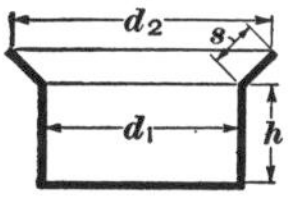	$\sqrt{d_1^2+4d_1h+2s(d_1+d_2)}$
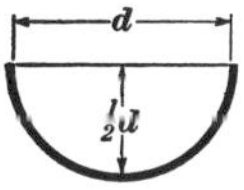	$\sqrt{2d^2}$
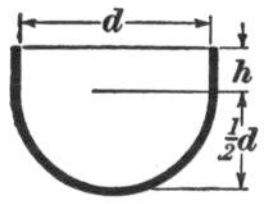	$\sqrt{2d^2+4dh}$
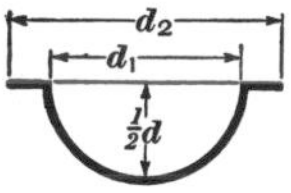	$\sqrt{d_1^2+d_2^2}$
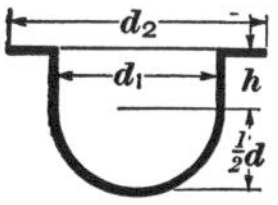	$\sqrt{d_1^2+d_2^2+4d_1h}$
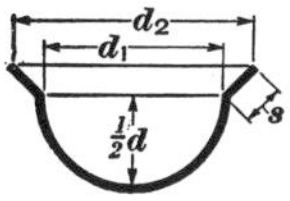	$\sqrt{2d_1^2+2s(d_1+d_2)}$

eter of a circle that has the same area. Of course, this method makes no allowance for thinning of the stock because of stretching, and so the diameter is only approximate. However, it gives a fair idea of the size of blank required.

72. Curling Dies. The edge of a drawn piece may be curled over to form a hollow ring by means of a special die. When this is to be done the blank must be cut to exact size and shape *before* forming, or trimmed to size after forming. A formed piece should be heat-treated before curling. The diameter of the curl, or rim, will depend on the thickness of the metal as well as its condition. Large curls should be avoided, for, if too large a rim is formed, the stretch of the metal will become excessive. The upper die *a*, Fig. 43, has a projection *b* that fits inside of the work, and a semi-circular groove in its face at the base of this projection.

TABLE I—(*Continued*)

Form of Shell	Diameter of Blank
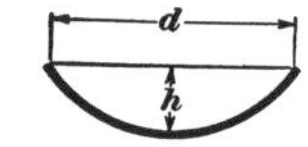	$\sqrt{d^2+4h^2}$
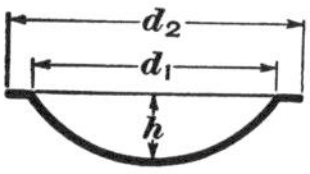	$\sqrt{d_2^2+4h^2}$
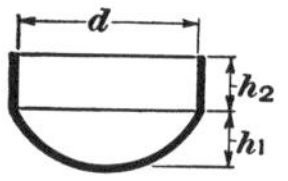	$\sqrt{d^2+4(h_1^2+dh_2)}$
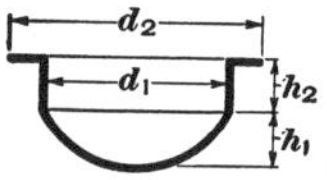	$\sqrt{d_2^2+4(h_1^2+d_1h_2)}$
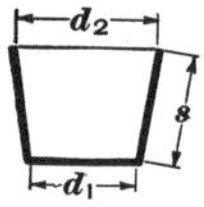	$\sqrt{d_1^2+2s(d_1+d_2)}$
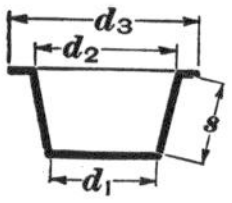	$\sqrt{d_1^2-d_2^2+d_3^2+2s(d_1+d_2)}$
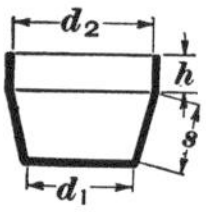	$\sqrt{d_1^2+2s(d_1+d_2)+4d_2h}$
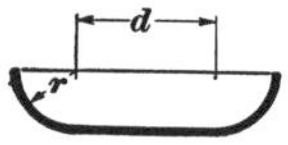	$\sqrt{d^2+6.3rd+8r^2}$
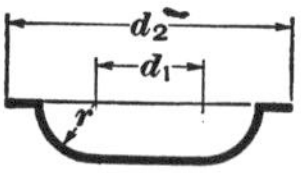	$\sqrt{d_2^2+2.3rd_1+4r^2}$
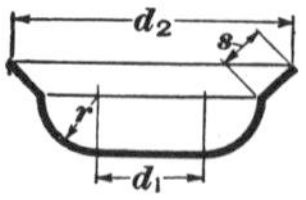	$\sqrt{d_1^2+6.3rd_1+4r(2r+s)+2s(d_1+d_2)}$

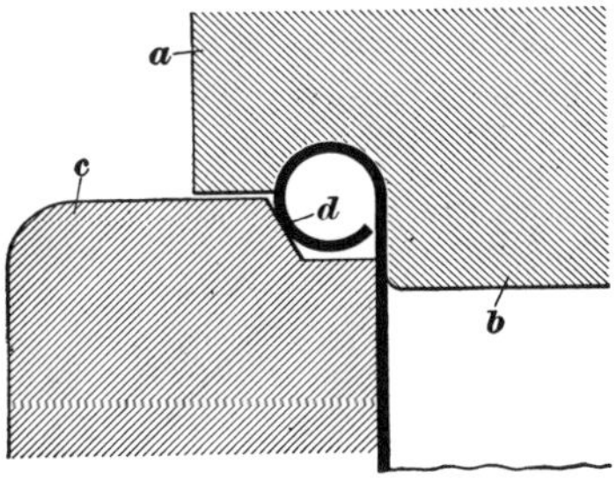

Fig. 43

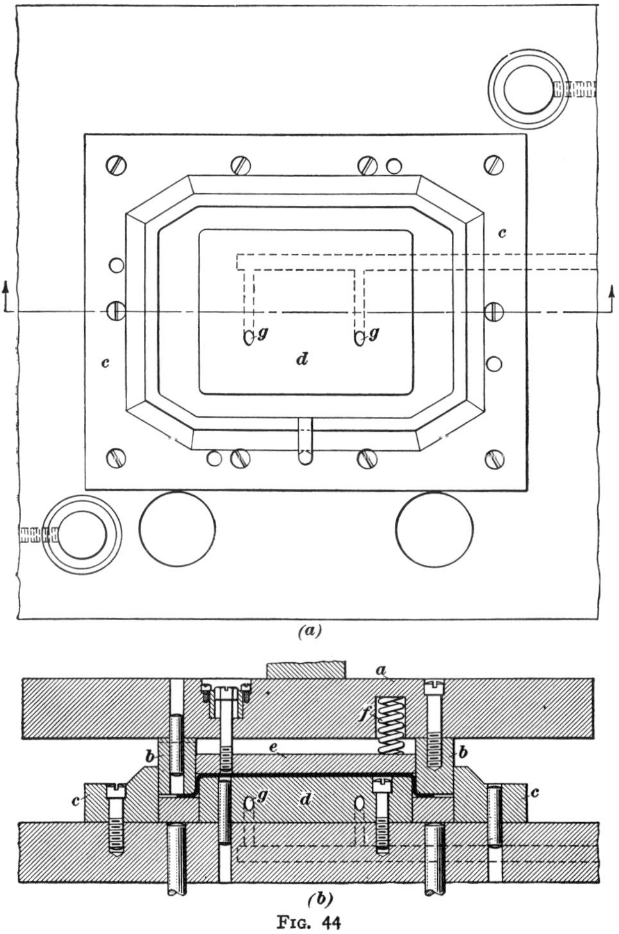

Fig. 44

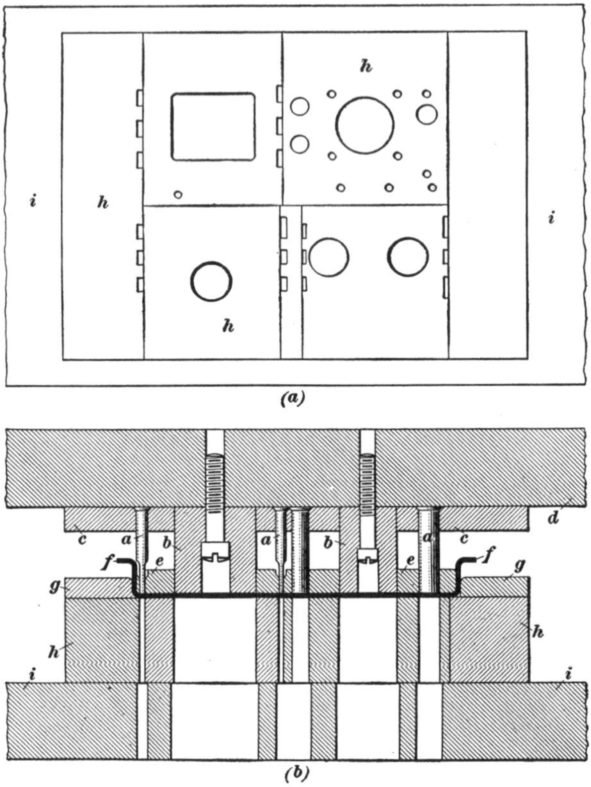

FIG. 45

The upper end of the lower die *c* is recessed at an angle of 60° to receive the rim as it is formed. As the upper die descends, the open end of the piece *d* is forced outward and follows the curvature of the semicircular groove. Because of the stretch given the metal, the circular form is retained by the metal after it has left the groove. The angle of the recess causes the circular section to backform against the wall of the formed piece. In some work, the edge is curled over a wire ring, in which case the curling dies employed are called wiring dies.

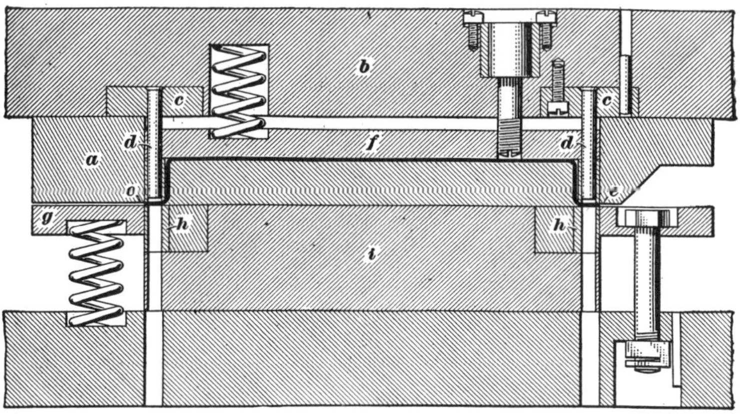

FIG. 46

73. Blanking-Drawing-Piercing Dies.—A type of die much in use on crank-presses is one known as the blanking-drawing-piercing die, by which several operations are performed, either progressively in a series of dies or at the same stroke of the press in a compound die. An example of the progressive type of dies is shown in Figs. 44 to 47, inclusive, in which four dies are used to blank the piece, to pierce a series of holes in the bottom of the formed piece, to trim the flange and to pierce a number of holes in it, and finally to pierce holes in the end of the formed piece.

74. The punch holder *a*, shown in the sectional view (*b*), Fig. 44, carries the punch *b*, which, in its descent, enters the die *c* and shears out the rectangular blank. Continued descent of the punch then forces the edge of the blank down over the central block *d* of the die and forms the turned edge and the flange, as illustrated. The die *c* and the block *d* are also shown in the plan view (*a*) of the die. During this operation, the blank holder *e*, view (*b*), under pressure of the springs *f*, clamps the blank against the block *d*. As the punch recedes, the holder *e* knocks out the formed blank. Vents *g* prevent the trapping of air.

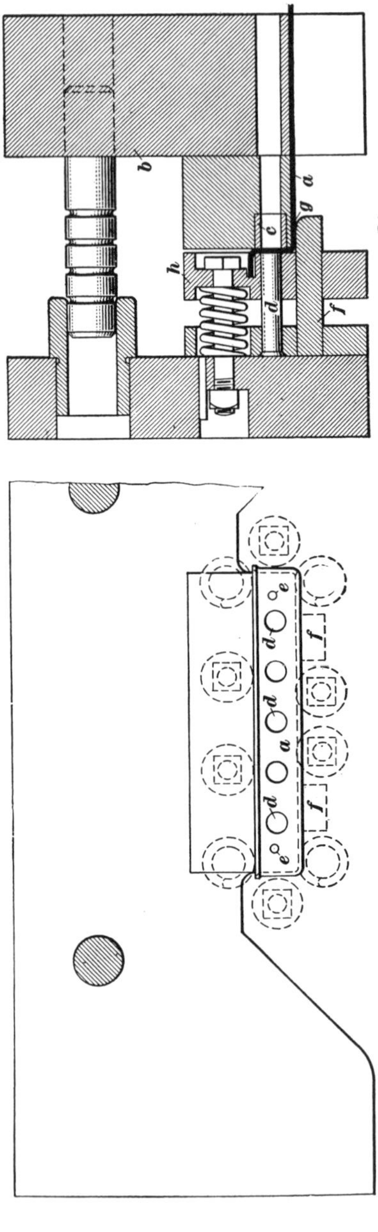

Fig. 47

75. The second operation is the piercing of the holes in the bottom, and this is done by the die set shown in Fig. 45 (*a*) and (*b*). The several punches *a*, *b*, etc. are fixed in the punch plate *c* attached to the punch holder *d*, and at their lower ends pass through the stripper plate *e*, which serves to guide them. The formed blank *f* is dropped between the gages *g*, and on the downward stroke of the punches the holes are formed. The die is shown at *h* and the die holder at *i*.

76. Trimming the flange and piercing the holes in it together constitute the third operation on the base, and Fig. 46 illustrates the punch and die for this operation. The trimming punch *a* is carried by the punch holder *b*, into which are set the circular plates *c* that hold the round punches *d*. The lower edge *e* of the trimming punch is below the lower ends of the round punches *d*, and so the descent of the punch holder first shears off the flanges of the piece and then pierces the holes in the flanges. The plate *f* acts as a blank holder and a stripper plate and the plate *g* is a stripper plate for the strap formed by trimming the flanges. Tool-steel inserts *h* in the die block *i* form the dies for the four circular punches *d*.

77. The final operation on the base is the piercing of the holes in the end, and this is accomplished by the die illustrated in Fig. 47 (*a*) and (*b*). View (*a*) is a plan view of the die, and view (*b*) is an end section. The base *a* is set over the die block *b* in a vertical position, which brings the end horizontal over the bushings *c* and under the punches *d* and *e*. To keep the base in position during the piercing operation, the punch plate carries two clamps *f*, which are short bars rounded on the inside lower edges. On the downward stroke, these clamps slide over the bottom of the base and hold it firmly against the die block, as indicated at *g* in view (*b*). The spring loaded stripper plate *h* removes the pierced work from the punches on the return stroke.

78. Compound Die.—An example of a series of operations carried out on a compound die is illustrated in Figs. 48 to 50, inclusive. The operations involved are as follows: The

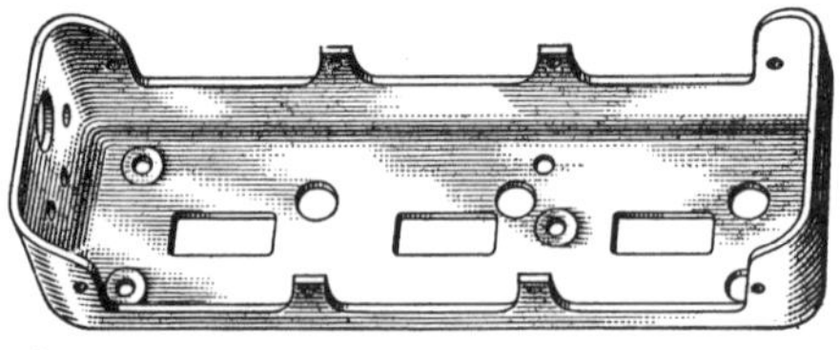

FIG. 48

blank is made from a piece that has been cut roughly to size, and this blank is brought to the required form by a drawing and embossing operation. Next the required holes are pierced in the bottom, and then a final piercing operation produces the holes in the sides and ends. The form of the finished piece is shown in Fig. 48.

79. It will be assumed that the blanks have been cut and are ready to be formed. The second stage, which brings the blank to the required shape, involves a drawing operation to form the flanges and an embossing operation to produce the circular indentations in the bottom.

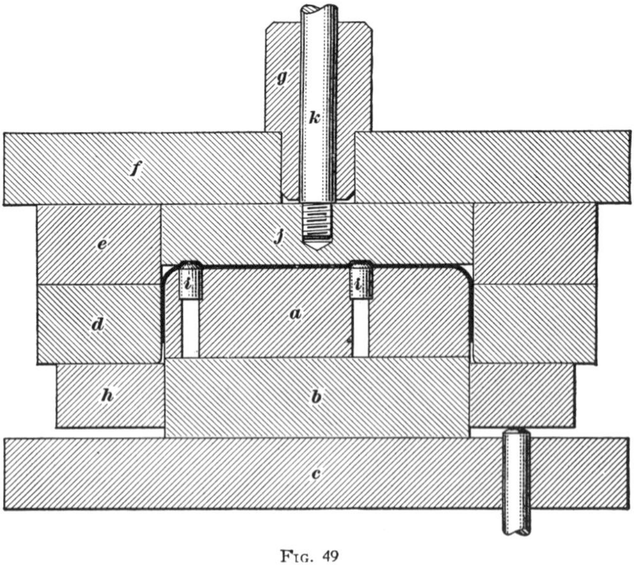

FIG. 49

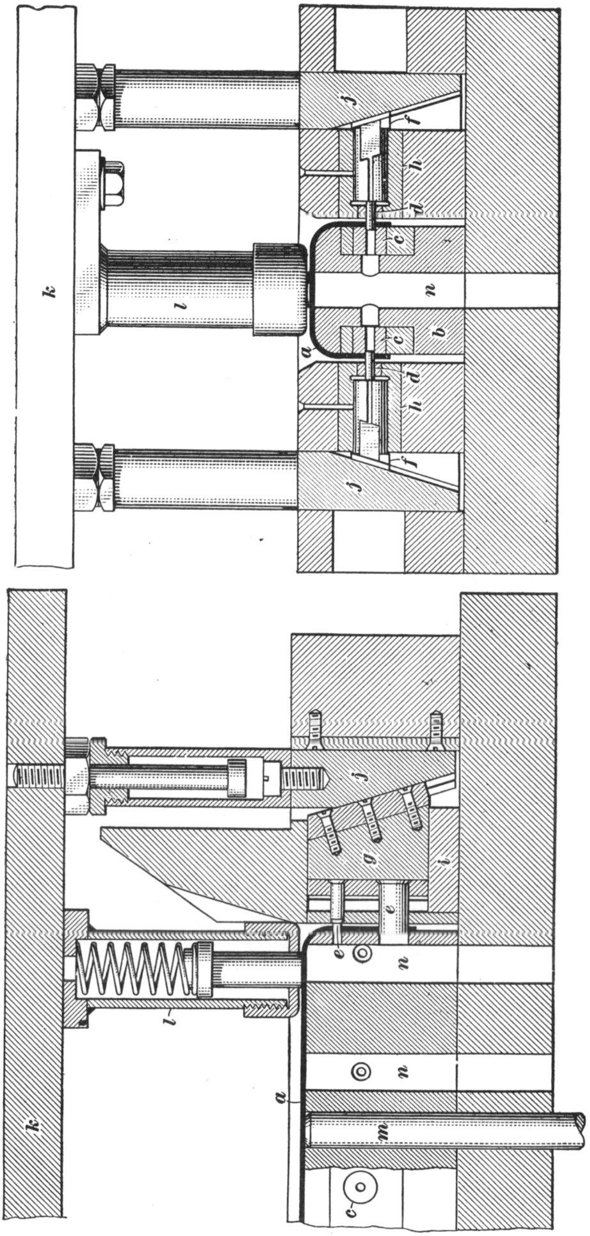

FIG. 50

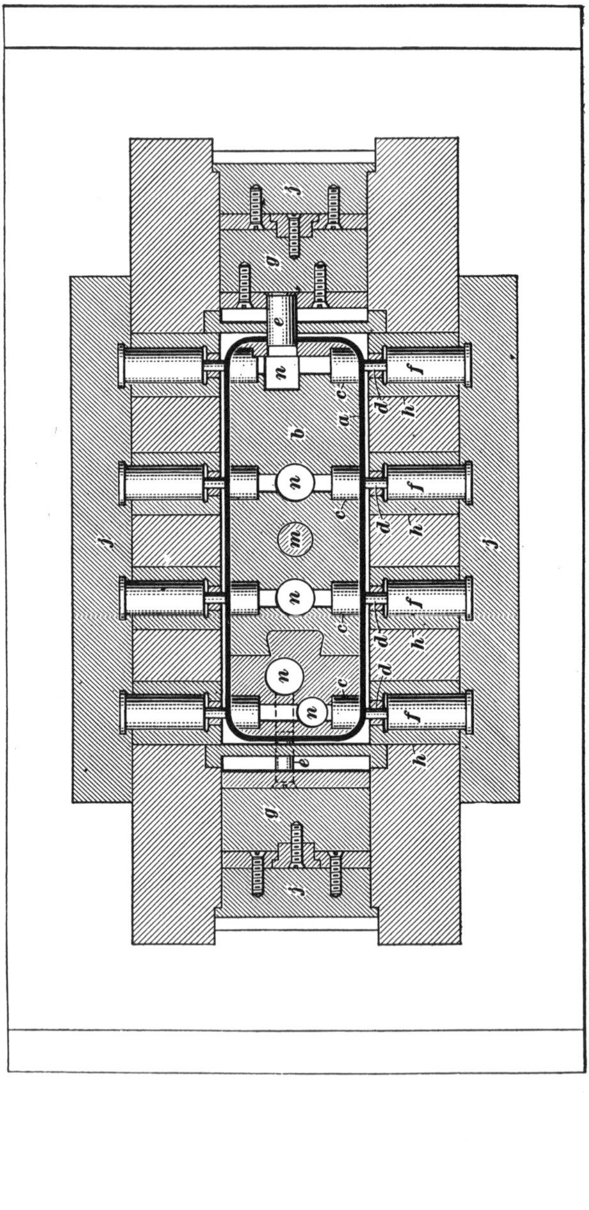

FIG. 51

A vertical section taken lengthwise of the die used for the combined drawing and embossing operation is given in Fig. 49. The draw punch *a* in this case is stationary, being supported by the space block *b*, which in turn rests on the bottom block *c*. The drawing die *d* is fastened to a pad *e* carried by the holder *f*, the shank *g* of which is copper-brazed in place. When the die is open, the spring-supported clamping pad *h* is flush with the upper surface of the draw punch *a*. The blank is placed on the pad and the press is tripped. The descent of the ram grips the edges of the blank between the die *d* and the clamping pad *h*, and the flanges are drawn down over the sides of the punch *a*. The inserts *i* produce the circular bosses, corresponding recesses being cut in the pad *j*. When the ram rises, the formed piece is ejected by pressing on the rod *k* attached to the knockout pad *j*.

80. Piercing the required holes in the bottom of the formed piece is a similar operation to that previously described; therefore, the punches and dies required are not shown. However, the piercing of the holes in the side and end flanges requires a special arrangement of punches and dies, as illustrated in Figs. 50 and 51, Fig. 50 (*a*) being a part front vertical section, (*b*) an end section, and Fig. 51 a horizontal section. The same parts that appear in more than one view are given similar reference letters, so that all three views should be referred to in reading the description.

The pierced frame *a* from the preceding operation is dropped over the die block *b*, which contains the dies *c* inserted in its sides. The punches *d* and *e* are carried by plungers *f* and *g* that are arranged to slide in suitable guide bushings *h* or on the block *i*. These plungers have inclined faces at the outer ends to match similar faces on the cams *j* attached to the top holder *k*. When the top holder descends, the cams force the plungers inwards and the punches are driven through the flanges of the frame *a*, piercing the holes. The spring-loaded plungers *l* hold the frame in position during the operation and the knockout bar *m* removes the frame after the operation. The punchings fall out through the openings *n* in the die block.

Fig. 52

EXAMPLES OF CRANK-PRESS WORK

81. Forming.—Short channel sections *a*, Fig. 52, may be formed to advantage on a hydraulic press using a special die in which a spring-loaded pressure plate is located. The die *b* is clamped to the press bed and the punch *c* is attached to the ram. A blank is located across the supporting faces *d* of the die, and is located by pins *e* in the pressure plate *f* lying between them. Holes corresponding to these pins are punched in the blank when it is stamped out. The punch forces the pressure plate downward and carries the blank to the bottom of the die, which is curved to the exact shape required. The radius of the punch is equal to that of the die less the thickness of the metal. As the punch rises, the spring of the pressure plate forces the formed piece upward and out of the die.

82. The piece *a*, Fig. 53, can be bent from the blanks *b* in a single operation by means of a die using a spring-loaded pressure plate and a punch having the exact shape of the required piece. A templet *c* is used to check the size and shape of the blanks if and when the operator of the press suspects that they are not perfect. A blank is laid on the faces *d* of the die, in which it is located by four steps *e* that form a part of the die. The punch *f* holds the blank firmly to the pressure plate *g* as it forces the blank into the die, the spring pressure being sufficient to prevent any slippage of the blank during bending. A stop *h* attached to one of the spring-loaded pins *i* of the die holds the pressure plate level with the face of the die when in its raised position.

83. Some pieces can be formed on a crank press to advantage by using special dies having spring-loaded pressure plates in both the die and the punch. In the case illustrated in Figs. 54 and 55, the finished piece will have a raised center section and a flange around the outer edge. A blank *a*, Fig. 54, is located on the die *b* either by pins or stops, and is held firmly

FIG. 53

in place by a spring-loaded pressure plate *c* at the center of the punch *d,* as the punch descends.

A further downward movement of the punch forces the blank against a spring-loaded pressure plate *e* located in the die, and draws the blank to the shape of the die. To avoid wrinkling, a notch *f,* Fig. 55, must be cut out of the blank *a* so that the flange can be folded up on both sides of the angle formed at this point. It is necessary to use the pressure plate *c* to prevent too free a movement of the center section of the blank, and pressure

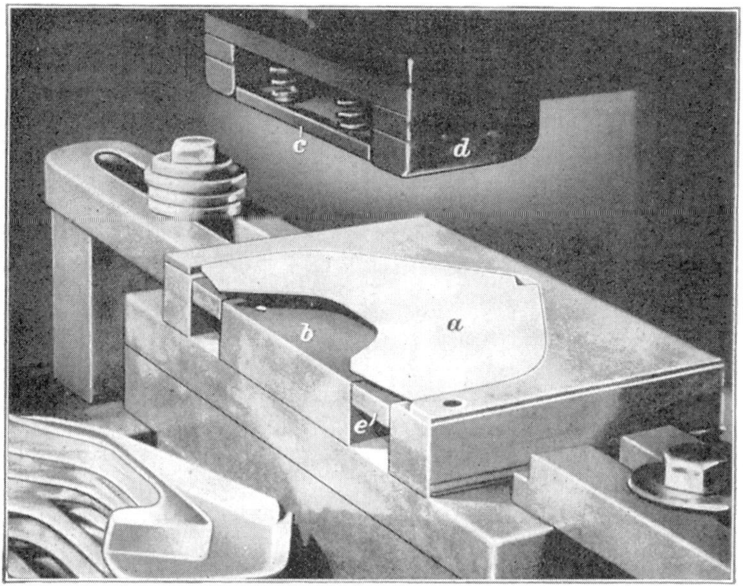

FIG. 54

between the punch *d* and the pressure plate *e* causes the blank to form tightly against the punch, which must therefore have the exact form required.

84. By means of progressive dies, two or more separate bending operations can frequently be carried out at one stroke of the press. The piece to be bent is moved from one part of the die to another following a progressive series of steps that deliver the part in its finished condition. A die of this type is shown in Fig. 56, in which the finished part is produced in two steps. A blank *a* is laid across the right-hand end of the die *b*, where it is positioned by pins *c* that pass through holes already punched in the blank. As soon as the punch *d* exerts pressure on the blank it bends, releasing itself from the pins, and the blank can then be drawn down into the die in which it is bent sharply to the required form.

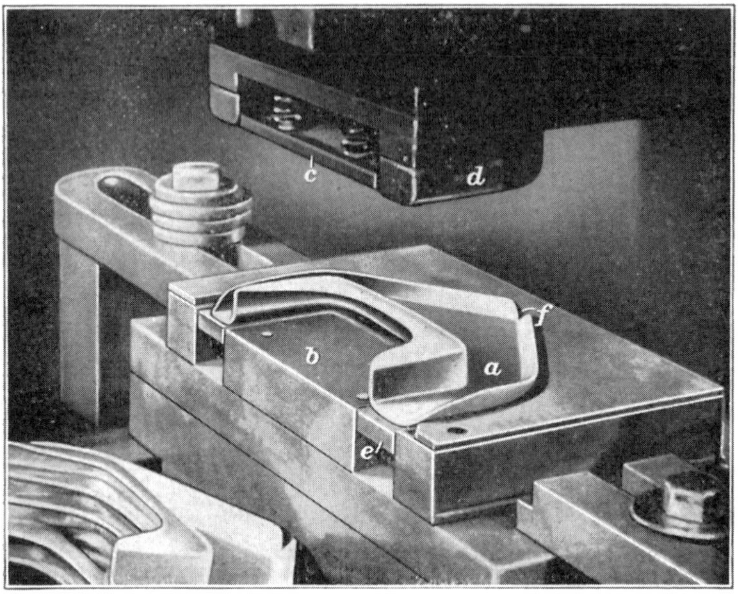

FIG. 55

85. The partly-formed piece is then transferred to the left-hand end of the die and slid over an arm or mandrel *e*, Fig. 56, which has six faces. The free end of the piece *f* is forced downward over the mandrel by the punch *g* to lie against the end *h* that was previously formed. The finished clip must be bent accurately so that the holes *i* in the ends of the clip will be in line. The arm *e* is spring-suspended and is normally in an open position, so that the piece may be placed in position or removed freely. The action of the punch *g* is to force the arm against the base of the die and to hold the formed portion in its correct shape as the free end is bent.

86. The piece *a*, Fig. 57, is formed from the blank *b* on a double-action die using a forming ring. When set upon a crank press the forming ring, or pressure pad *c*, is flush with the top of the punch *d*, being supported on an air cushion in the base of the press. As the punch descends, carrying the die *e* with it,

FIG. 56

a spring-loaded knockout *f* located in the die presses on the punch and helps to hold the blank in place. Pins located in the punch serve to locate the blank. As pressure is exerted on the blank the pressure pad is forced downward on the air cushion, and the slippage between the die and the pressure pad permits the blank to form over the punch. At the same time the spring-loaded knockout *f* is forced back into a recess on the die. As the knockout is forced into the die it punches a hole *g* in the blank and forms the edge of the metal surrounding the hole. A punch located in the die passes through the knockout and enters the hole *h* in the punch *d* in order to punch out the smaller hole in the formed piece. The spring load on the knockout is much less than that of the air cushion, so there is no tendency for the piece to reform as the ram is raised.

87. Joggles.—It is frequently necessary to form an offset, or joggle, in a stringer piece where two such pieces cross, or

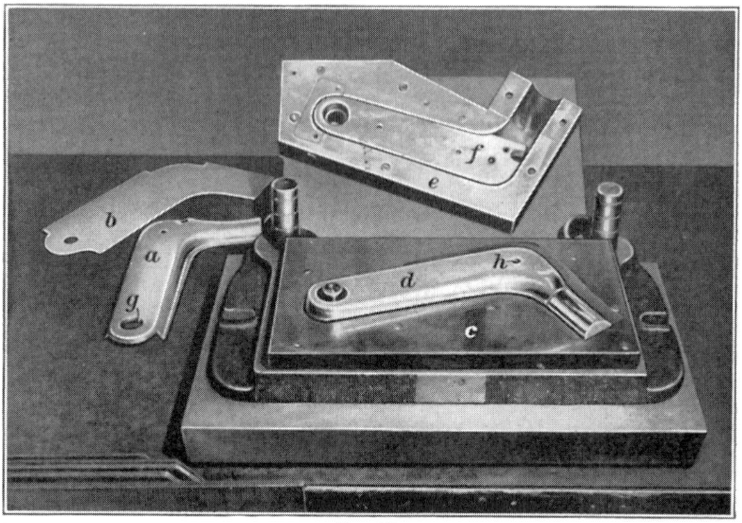

FIG. 57

where the ends overlap in order that the skin sheet may be smooth and lie in close contact with the stringers at all points. Such offsets can be formed in a crank-press by the use of a joggle die such as that shown in Fig. 58. A joggle is required at the end of the bulb section stringer *a,* which makes necessary the deforming of both sides of the section. The side *b* on which the bulb is located is formed, or offset, by keeping the plane of that side true and by forcing the joggled section downward in the original plane. The side *c* is bent downward, keeping the plane of the joggled section parallel to that of the stringer.

88. The lower part of the die is made in three sections, the base *d,* Fig. 58, that is attached to the press bed, a forming block *e* that is seated solidly in the base, and the joggle section *f* that rests on a rubber pad *g* and moves downward under pressure. The upper part of the die consists of a holder *h* attached to the ram, and two die blocks *i* and *j* held firmly in the holder. The amount of the joggle given the stringer depends on the distance that the block *j* extends beyond the die holder as com-

FIG. 58

pared to the block *i*. A space *k* between the blocks *i* and *j* determines the sharpness of the joggle. As the ram descends, the block *j* presses against the stringer *a* and forces the block *f* down, compressing the rubber pad until the block *i* makes contact with the end *l*, bending it to its true position. Both blocks *i* and *j* must be given an offset *m* on their rear faces, in order to clear the bulb as the dies descend. Some form of stop is provided by which the stringer can be definitely located in the dies. This stop may be in the form of a pin set in or clamped to the die block *e*, or may be a piece of adhesive tape stuck on the inner faces of the V.

89. Stamping.—The crank press finds its greatest application in stamping operations. The templet *a*, Fig. 59, shows that the corners of the blank *b* must be removed in order that the piece may bend into the form of a shallow box, or fit into the position for which it is intended, as the case may be. Guides *c* and *d* are bolted to the press-bed table and are accurately set so as to locate the blank under the punch *e*. Only one corner of the blank *f* can be stamped out for each setting of the guides.

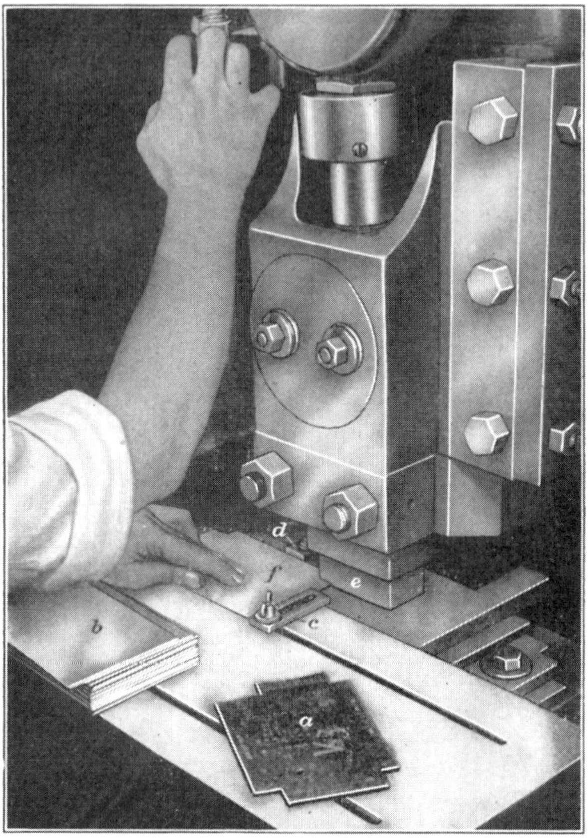

FIG. 59

When that corner has been removed from all of the blanks the guides are reset for the next operation. A new setting is made for each corner. To set the guides the punch is lowered, the templet is placed against it, and the guides are bolted in position. The stamped pieces should be checked against the templet from time to time to insure against any shift of the guides.

CAST AND MOLDED DIES

METAL DIES

DESIGN CHARACTERISTICS OF DIES

1. **Nature of Press Dies.**—A press die is a tool for cutting or forming metal or other materials. Dies are generally made up of two components—the die proper into which the material is forced and brought to the required form or shape, and the punch by which the metal is forced into the die. Since the cavity into which the metal is forced is formed in the die, this is known as the female part, while the punch which acts on the material is known as the male part. Either part may be set above the other in the machine but it is more usual to place the female part, or die, on the machine bed and attach the punch, or male part, to the moving head. The punch may be formed to the exact shape of the die or may be partially formed and some medium such as a rubber pad used to force the metal to the required shape. The material to be worked is generally in the form of sheets or long strips. During the die action the metal may be cut, bent, punched, crowned, stretched or may be flowed according to the nature of the required part.

2. Dies must be designed for the particular metal to be formed or cut. Consequently, those properties of the metal which affect its action during forming or cutting must be understood. These properties are referred to as strength, elasticity, ductility, and hardness, all of which can be held reasonably constant by modern metallurgical processes. Two other properties that affect the action of the metal are the ultimate strength at which the metal fails utterly and the elastic limit at which permanent deformation takes place.

Ductility is the property of a metal by which it will undergo plastic deformation without fracture. All of these properties are interrelated and a control of them all is best obtained, for any metal, by controlling the degree of hardness developed. Hardness of a metal may be regulated by controlled heat treating or by cold-working it.

In the case of aluminum alloys, however, there is likely to be sufficient variation in stock supplied at different times or from different mills to warrant close attention to the action of the sheets or strips as production begins on any particular job.

3. **Work Performed.**—The three distinct, or basic, actions in the work performed by dies are cutting, forming, and pressing. Each of the three requires dies especially designed for the purpose intended and these dies must also be planned for the machine best suited to the type of work involved. In some cases the same general type of die may be used on different machines, as on either the drop hammer or the hydraulic press, with only minor changes; but the better practice is to design for the machine on which the work is to be done.

4. **Cutting.**—By the mating of the punch and die, a cutting action is produced on the metal placed between them. Both punch and die must have sharp edges. The shearing action that takes placc is, for all practical purposes, simultaneous with the application of force and is confined to the area between the vertical planes passing through the acting edges of the punch and the die. The width of this area is equal to the clearance of the mating parts and the amount of the clearance will depend on the nature of the metal being cut. For dural the clearance will vary from 8 per cent to 14 percent of the thickness of the sheet, the exact clearance for any die being based on past experience with the shape of the die involved and the metal to be used.

5. Clearance is obtained by altering the dimensions of either the punch or the die from the true part dimension. Generally, the clearance is obtained by altering the dimensions of the punch as in the case of blanking. In other cases, however, as for piercing operations, the clearance is taken from the die. This method keeps the blank within the true part size and insures a hole no smaller than that called for. In addition to the clearance, some back draft, or taper away from the shearing edges, is given

both the die and the punch. This prevents binding between the parts as the metal is sheared and also lessens the pressure between the blank and the die after the punch has entered, or telescoped into, the die. A 3 per cent taper is generally used. While it is usual to apply the relief to the die, as in the case of a cutting die, some cases, such as a continental die, require that the taper be given to both the punch and the die.

6. Cutting may take place along the entire periphery, or outline, of the die at the same instant or may be a progressive action, depending on the shape of the piece required. Blanking dies are usually so made that the entire cutting action takes place instantaneously, whereas punching dies may be given a sloping cutting edge that will produce progressive cutting. A shearing action is then produced. The selection of the machine on which the work is to be done will largely depend on the pressure required to shear, or cut, the metal. This pressure must be determined by the formula

$$P = slt$$

in which P = pressure, in pounds;
s = tensile strength of the metal, in pounds per square inch;
l = length of cut, or periphery of the die, in inches;
t = thickness of metal, in inches.

Any machine capable of cutting out the entire periphery of the die will, of course, have more than sufficient power for progressive cutting.

7. Progressive cutting, or shearing, may be calculated on the basis of a modification of the formula for cutting by using a factor f that provides for either full shear or for half shear. No advantage is gained by less than one-half shear and there is no advantage in more than full shear. The formula then becomes $P = sltf$, in which the value of f varies from $\frac{1}{4}$ for full shear to $\frac{1}{2}$ for half shear.

8. Forming.—A forming die permanently alters the shape or contour of a flat piece of sheet metal or extruded section and may require either bending or drawing of the metal, or both.

Forming dies may be used on punch presses, power brakes, drop hammers, or hydraulic presses. In any case, the elastic limit of the metal must be exceeded to give the internal movement, or flow, in the metal necessary to a permanent set. Bending operations usually require less actual flow of the metal than draw forming.

9. The simplest type of die forming is the bending of a sheet to a radius along a straight line. A forming die, however, may produce bending along more than one such line provided that the bends do not cross or interfere with each other. Bending produces a compression on the inside of the curve and a stretching on the outside, the two actions starting from the neutral plane of the sheet. The two actions balance, or compensate each other to a large extent, depending on the compression and stretching qualities of the metal; but there is generally a slight reduction in the thickness of the metal, as well as stretching, or increase in length, as the bend progresses. The allowance for a bend is found by the formula

$$A = (.0078\,t + .01743\,r)n$$

in which A = allowance, in inches;
t = thickness of blank, in inches;
r = inside radius of bend, in inches;
n = number of degrees of bend.

10. Drawing.—Both compressing and stretching of the blank are involved in drawing since the flat blank must first be reduced in area as it is drawn over the radius of the die into the die cavity and must then be stretched to conform to the irregularities of the die itself. The result is an extensive flow of the metal during the drawing action with a resulting tendency to change the thickness of the metal. The flow of the metal may take place in a number of directions, depending on the shape of the die, and it is important that this flow be controlled in order to prevent the formation of wrinkles that may be drawn into the die or an unevenness in the surface of the metal. A perfect drawing action should result in a total area and thickness equal to that of the original blank but in practice this ratio will vary slightly.

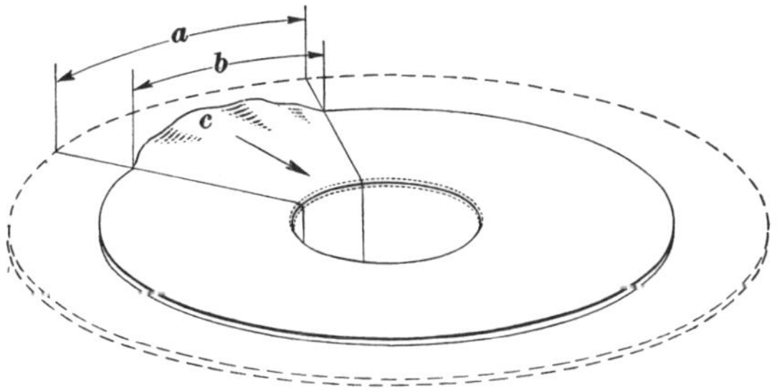

FIG. 1

11. The compressive action that takes place during drawing is illustrated in Fig. 1. Any segment *a* of the outer edge of the blank may be taken as a test section of the blank. During the drawing process the center of the blank is drawn down into the cavity of the die and at the same time the outer edge of the blank is drawn toward the center a corresponding distance. At the position shown, the length of the arc *a* must be reduced to the corresponding length *b*, since the test section will be bounded by radial lines emanating from the center of the blank.

12. Aluminum and its alloys have a low compression factor, and consequently, unless the outer edge of the blank is restrained in some way that will force the compression of that portion of the blank lying outside of the die, a wrinkle *c* will be formed, as shown in Fig. 1. If unrestrained, the wrinkle may even extend down into the die where it must be ironed out by the punch at the expense of extra power or by a separate operation after the piece is formed. The working pressure required by forming operations is found by the formula

$$P = 3.1416\,dts$$

in which P = working pressure, in pounds;
d = main diameter of shell, in inches;
t = thickness of stock, in inches;
s = tensile strength of stock, in pounds per square inch.

13. Forming Data.—The three principal factors involved in metal forming are: the movement of the metal, the pressure required, and the load limits. All formulas developed for forming operations will include certain intangible factors which in the final analysis depend on the personal equation of the operator and which, therefore, cannot be calculated. As a result, most formulas are reduced to their simplest form and the experience of the operator is depended on to establish the remaining factors. Also, the relatively wide variation in materials prohibits too much dependence on highly theoretical formulas.

14. Energy, Impact, and Pressure.—The rate at which a metal will flow at any given temperature limits the amount of deformation of that metal for any one operation and, therefore, limits the forging, or forming, power that can be applied. If the metal is deformed beyond the rate of plastic flow, rupture of the metal will result. The rate of plastic flow will vary with the kind of metal involved, its form and density, the temperature of the metal during the operation, the speed, or velocity, at which deformation occurs, and the continuity of the forming action.

15. The forging energy required to produce a part is found by the formula

$$E = \frac{w\,v^2}{2\,g}$$

in which E = energy required, in foot-pounds;
w = weight or force applied, in pounds;
v = velocity of forging element at point of impact, in feet per second;
g = acceleration due to gravity, or 32.16 feet per second per second.

Assuming a freely falling body such as the head of a drop hammer, as the basis of analysis, the weight becomes the only variable, indicating that the weight or pressure exerted must be greater than the affected mass of the piece being formed.

16. The applied force must first overcome the resistance to deformation of the metal and must then produce the required flow of metal. Once started, the flow of the metal will continue as long as the applied force continues in motion and in magnitude. With continued application of the applied force, the density of the metal is increased, causing an increase in the resistance to flow. Also, the friction between the part and the forming, or forging, dies will increase with the magnitude of the applied force.

17. Plastic deformation, or forging, may be accomplished either by impact, as in the case of the drop hammer, or by pressure, as in the case of the hydraulic or crank press. Deformation by impact makes use of the maximum force during the period of contact between the work and the dies and the forging effect continues until the increasing resistance to flow brings the die to rest. All of the useful energy of the machine is absorbed. Once the flow of metal is started, it acquires a certain momentum, and a greater rate of flow per blow can be obtained by rapid blows than by blows at greater intervals. The amount of forging, or deformation, that can be accomplished by pressure is limited by the original and increasing resistance to flow in the metal, the increased density of the metal between the dies, the friction between the work and the dies, and the change in temperature of the working during the operation.

18. Bending.—A certain allowance must be made for a bending operation when cutting the blank because of unequal stretching and compression of the material. The neutral plane about which the bend takes place does not lie at one half the thickness of the metal nor does the neutral plane follow a true arc. A true arc may be used for all practical purposes if allowance is made for this variation. Then, the length of the arc is

$$l=(r+.40t)\left(\frac{2\pi a}{360}\right) \qquad (1)$$

in which l = length of arc, in inches;
r = radius of inside curve, in inches;
t = thickness of metal, in inches;
a = angle through which metal is bent, in degrees.

The total length L of the blank will be equal to the sum of the lengths of the three straight sections b, c, and d, Fig. 2, plus the lengths of the two arcs l_1 and l_2, or

$$L=b+c+d+l_1+l_2 \qquad (2)$$

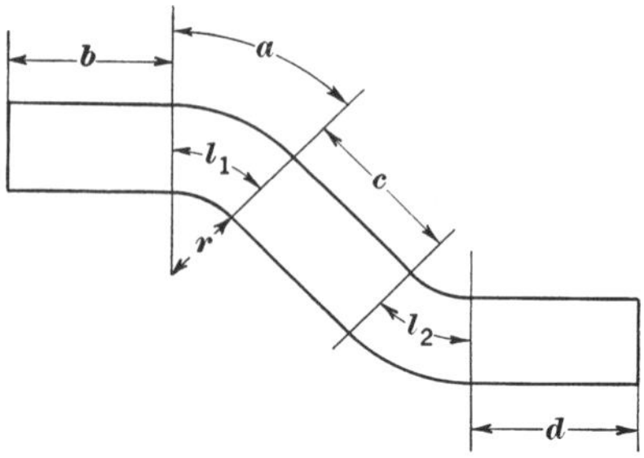

FIG. 2

19. Springback.—When aluminum alloys are bent or formed there will be some springback, or tendency to spring away from the formed shape, and this is especially pronounced in hard stock. To produce a part that is formed exactly to shape, the dies must either overform the piece by some means, or sufficient pressure must be exerted to set the part at the point of curvature. Because of the variable factors involved, the amount of overforming must be determined by experiment on the particular piece involved, or based on the known springback of similar parts.

20. Overforming may be accomplished by cupping the bottom and relieving the sides of the punch. This causes the part to overbend somewhat so that when the pressure is released the spring of the bottom will bring the sides to their true position. Another method is to recess the bottom of the die so that the forming pressure is exerted at the point of bend, in which case the concentration of pressure is sufficient to set the metal as the die bottoms. This is actually a coining operation and the compression strains set up balance or counteract the tensile surface strains.

21. Extreme care must be used in any bottoming procedure, as pressure builds up rapidly once the punch has bottomed, and any carelessness in adjusting the punch or in placing more than one blank at a time on the die may cause serious damage to the machine. The radius of the bead formed by recessing the bottom of the punch should be but little more than the thickness of the

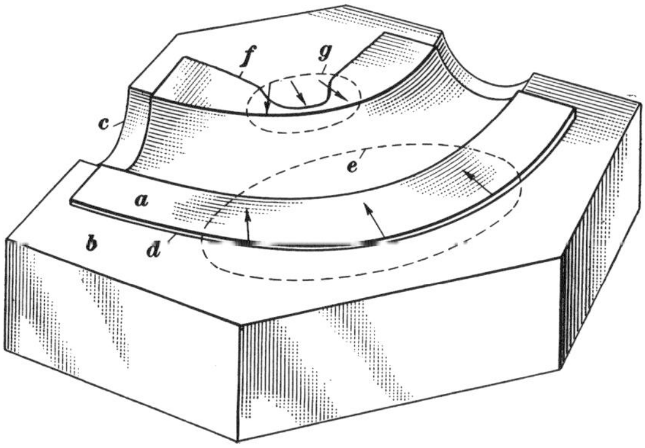

FIG. 3

metal. The same principle can be applied to dies for sharp-cornered V bends by constructing a bead at the point of the punch. In this case the bead draws the blank into the die and sets it at the bend without exerting pressure on the sides of the V.

22. Shrink and Stretch Areas.—The shrink and stretch points on a blank can best be illustrated by a die such as that shown in Fig. 3, in which the blank is formed to a 90-degree turn. A blank *a* is placed on the die *b* so that an equal amount of overlap is allowed on both sides of the die cavity *c*, and around the edge. During the forming operation the blank is forced into the cavity and some movement of the blank results. The outer edge *d* of the blank must be shrunk during this movement as the length of the arc decreases, and consequently the metal within the area *e* must be compressed, or wrinkles will result. The movement of the metal toward the cavity will be at a minimum at the ends and at a maximum at the middle of the arc, so that maximum compression or shrink will occur at the radial center line of the blank.

23. The distortion of the blank in Fig. 3 is more pronounced along the inner arc *f* since the action, or flow of metal, is confined to a smaller area *g*, and the tendency to stretch the metal in this area causes a greater slip of the blank across the face of the die. Except for sufficient slip at the ends of the blank, necessary to curve the metal into the cavity, very little distortion occurs; but along the radial center line of the blank the distortion reaches a maximum. In the area *g* the flow of metal is a combination of that due to the direct stretching of the metal into the cavity and a distortion of the blank as it is drawn over the edge of the cavity. The arrows indicate the direction in which the metal flows and in which the stretch and shrink occur.

24. An offset or joggle on one side of a formed piece or a sloping surface in the cavity would cause the blank to slip between the punch and the die instead of forming properly with the correct flange around the formed contour. To prevent this, lock rings or lock flanges are frequently formed on the die and a draw ring with a corresponding groove or recess is used to hold the blank in place during the operation. In the case of the die shown in Fig. 4, a joggle is to be formed at one end of the part and means must be provided to prevent or to control any movement of the blank. The required joggle *a* is formed on the die and a corresponding recess *b* is formed in the draw ring. The face *c* of the die is cut back, outside of the joggle, to the main plane of the die and the draw ring given a corresponding flange to form a lock flange so that the outer edge of the blank is held firmly during the operation. A sharp lock flange *d* will stop all slip of the blank but this flange can be modified to give a considerable slope *e* and allow whatever amount of slippage may be found necessary. The edge *f* of the opening in the draw ring should be almost directly above the forming edge of the cavity to prevent the formation of wrinkles in corners. That portion of the die face covered by the draw ring is known as the draw surface.

PATTERNS

25. Origin of Patterns.—All patterns from which dies are cast have their origin in the Engineering Department and on the loft boards. The pattern shop is furnished blueprints of the part to be formed and loft boards on which the part has been developed to full scale. Shrink templets are also furnished by the loft-

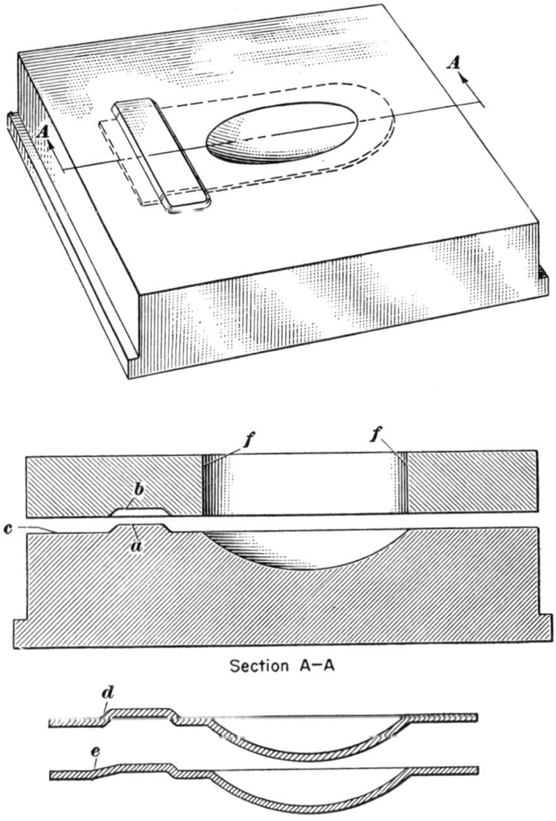

Fig. 4

ing department and these shrink templets may be used directly in the construction of a full-scale model, or mock-up, or may be used as positive templets from which matching or negative templets may be made to be used in checking the outside of contours. At this point the pattern maker must decide whether a wooden mock-up must be made or whether the required part is of such a nature that a plaster mock-up will be satisfactory. He must also decide on a method of procedure that will give satisfactory results with the least number of reproductions. Each time one pattern is made from another, as, for example, a negative pattern from a positive pattern, some error will be introduced because of the scraping and smoothing of the plaster in bringing it to the required finish.

26. The parts to be formed, and therefore the patterns, may be divided into two classes: Those which are critical in their shape and dimensions, that is, those which must fit exactly into other established sections of the airplane, such as the fuselage frame for a baggage door; and those which will allow for some slight deviation, such as a fairing strip. Parts that are critical require a pattern that is exact and consequently one that can be fitted to a test block having the required contour, and blued, called a blue block, whereas less critical parts need only conform to the requirements of the blueprint. The mock-up for a critical part may be made from wood or metal, neither of which is affected by the bluing material. Plaster should not be used for such a mock-up, since the bluing material works into the plaster and softens it. As a result, the plaster is easily deformed and will lose its correct contour.

27. The pattern maker must also decide whether to use a wooden mock-up or one made of plaster on the basis of costs and time involved relative to the number of parts required. A wooden mock-up is the more expensive of the two but its use may reduce the number of patterns required. The plaster mock-up is cheaper and more quickly made. However, a plaster mock-up requires additional plaster patterns in cases where the wooden mock-up could be used for making the sand mold from which either the die or the punch can be cast.

28. Patterns From Wooden Mock-Up.—A wooden mock-up that is small enough to be used as a pattern from which a sand mold can be made may be positive, that is, the outside of the mock-up may be the exact shape of the required part, or it may be negative, in which case it would have the shape of the die into which the part will be pressed while forming. Mock-ups are generally positive, however, since outside contours are more easily fashioned by wood-working tools. The casting made from

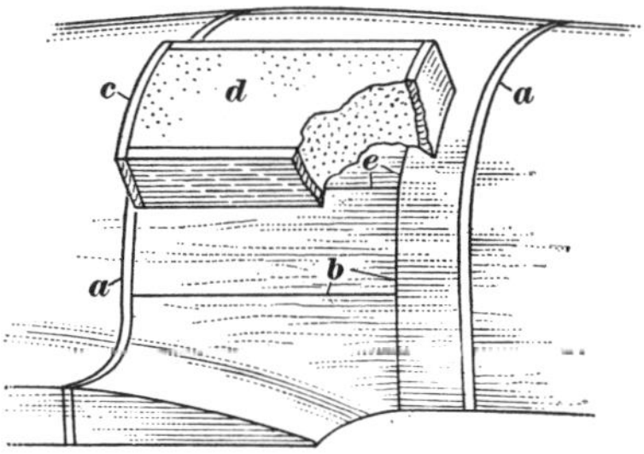

FIG. 5

a positive mock-up used as a pattern is the punch by which the part is formed. A plaster pattern is made from the wooden mock-up and is used as a negative pattern from which the die is cast.

29. A wooden mock-up that is too large to use as a pattern must be laid out in sections each of which is of such a size that a plaster pattern can be made from it. The outline of the section is scratched deeply into the wood and the scratch marks made sufficiently wide to permit of being reproduced in a plaster cast. An example of such a mock-up is shown in Fig. 5, which represents a portion of an airplane fuselage. The mock-up is made by first locating wooden templets *a* in their proper places and then filling in between these templets with wooden strips which are faired, or brought to their proper contour and flush with the templets. The section lines *b* are scribed into the surface and generally are identified by a narrow black stripe painted over them.

30. A wooden frame *c*, Fig. 5, is set over the section from which the negative pattern is to be taken and is filled with plaster. It is necessary that this frame shall be larger than the required section so that the plaster *d* will extend beyond the section lines *b*. The plaster is reinforced with hemp or other suitable material, and when hardened will represent both in contour and outline the area from which it was taken. This negative pattern may now be used as a mold from which a positive pattern may be cast to be used as a pattern for the punch.

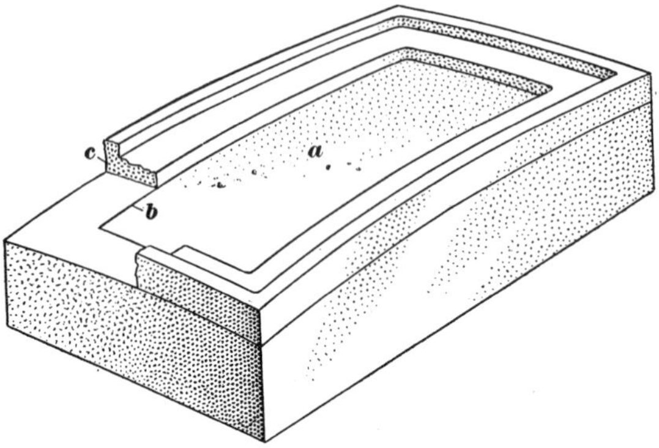

FIG. 6

31. No attempt is made to provide for flanging when this first negative pattern is made so that it becomes necessary to use the positive pattern as a mold from which to cast a second negative plaster pattern from which the metal die can be cast. In some cases the original negative plaster pattern can be reworked outside of the section lines so as to provide for flanging on the die and when this can be done the second negative pattern is unnecessary. When this is practicable it should be done, because each time one plaster pattern is made from another some accuracy of contour is lost. Also, part of the time required for making the second negative plaster cast is saved.

32. Patterns From Plaster Mock-Up.—Once a large plaster mock-up has been made, the procedure for producing the negative and positive patterns is very much the same as in the case of a large wooden mock-up. When small parts are involved, however, the original may be either a negative or a positive pattern, depending on which is the more easily made. When possible, a positive plaster pattern or mock-up is made, and from this a negative pattern is taken. Generally the negative pattern is used for making a sand mold, and a Kirksite die is cast in this. If a Kirksite punch is to be used, the positive pattern is also used for making a sand mold and the punch is cast; otherwise, the punch is cast on the Kirksite die.

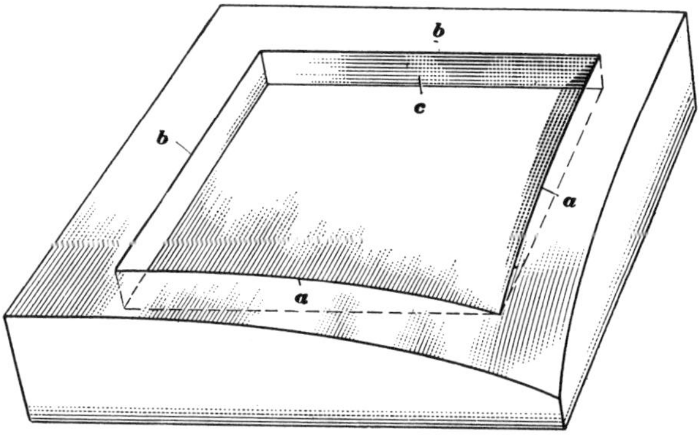

FIG. 7

33. When the shape of the section taken from a large mock-up permits, it will frequently save time and trouble to rework the negative plaster cast, or flash, originally taken so as to provide for flanging rather than to make a positive and then a second negative. This is especially true when by so doing the positive plaster pattern can be eliminated. The block *a*, Fig. 6, represents a negative plaster casting as taken from a section of a large mock-up, and the line *b* indicates the area of the section. After the pattern has been removed from the mock-up, a section *c* is added to provide for flanging. A sand mold is then made of this negative pattern from which the die is cast. It will be unnecessary to use a positive pattern, as a lead punch can then be cast directly on the Kirksite die.

34. An example of a critical die used for forming the baggage door fitted to the nose section of a plane is shown in Fig. 7. This door is located above the center line of the nose section and consequently presents a warped surface, since the forward and upper lines *a* of the door frame will curve much more sharply than the aft and lower lines *b*. The flanges *c* of the door opening must be in line with the direction of the hammer-head movement and consequently the back of the die must lie in a horizontal position. For this reason one end of the die will be much thicker than the other. Unequal cooling will take place throughout the mass and consequently unequal shrinkage will occur. Since the frame must fit exactly to the nose section, this unequal shrinkage must be provided for when the pattern is made and the die must be made to a blue block, or a check grid, made from the master mock-up.

CAST DIES

35. Material Requirements.—Drop-hammer dies may be made from pure zinc, zinc alloy, or aluminum alloy scrap, but the most satisfactory metal for the purpose has proved to be a zinc alloy known as Kirksite. Dies used in a drop hammer are subjected to severe punishment and must be able to withstand the impact of the punch without excessive distortion and without cracking or breaking. Because of the relatively small number of pieces to be made from any one die, the cost of producing the die, including the man hours involved, becomes an important factor; and because of design changes which may be introduced the dies should be of such a nature that they can be produced with the least possible delay in production. This condition practically excludes such metals as cast iron from the list of possible materials because of the time required to produce the die and the skill required in bringing such a die to the required contour and finish. Dies can be made best from a metal having a low melting point and relatively light weight and which, while hard enough to withstand the required service, still will be soft enough to work easily.

36. Zinc.—Dies cast from zinc have proved satisfactory within certain limits but are not tough enough to withstand the severe usage of a relatively long run. The cost of production is low and castings are readily made. Very little handling of the cast die is necessary to prepare a die cast from zinc for production work and such finishing as may be required can be done by means of a rotary file, or burr, when metal must be removed, and by emery cloth disks and drums for finishing and polishing. When pure zinc is used, a pouring temperature of 820° F. is satisfactory; but as the amount of impurities contained increases, this temperature will increase to as much as 860° F., which is the maximum temperature that should be used. The tensile strength of unalloyed zinc is about 600 pounds per square inch (p.s.i.) when new, but decreases after remelting a few times.

37. Foreign matter or impurities in the zinc will increase the rapidity with which the metal will chill during pouring, so that the higher temperatures must be used to insure a truly molten condition of all metal entering the mold from which the die is cast. The impurities in the zinc are removed by the addition of sal ammoniac, which causes the dross to rise to the surface of the molten metal from which it can then be skimmed off.

38. Zinc dies have not proved satisfactory for forming in the drop hammer for either long runs of heavy metal or for stainless steel. Distortion soon becomes excessive under the severe usage necessary to forming heavy metal and particles of the zinc die adhere to the stainless steel. During heat treating these particles amalgamate with the stainless steel and cracks develop at the point of fusion.

39. Kirksite.—Zinc having a purity of 99.99 per cent, mixed with small quantities of aluminum, copper, and magnesium, gives an alloy which has all of the qualities of zinc for making cast dies plus additional qualities that overcome the objections to that metal. Sand molds have proved very satisfactory for casting dies from Kirksite. About 94 per cent of Kirksite is zinc. A Kirksite die may be used with a lead or zinc punch or with a punch made of Kirksite. In any case the punch may be cast directly in the die, eliminating the necessity of making a special mold for the punch. Kirksite is relatively light in weight, melts at low temperature, about 717° F., and is easily worked. It can be brought to a high mirror-like finish. Dies made from Kirksite can be remelted when they have served their purpose, but this metal is likely to contain impurities, especially if a lead punch is used.

40. The presence of lead in Kirksite is hard to detect and it is difficult to remove the lead from the alloy. Lead contamination may not be suspected until it is found that the die will not stand up to the work required of it, and will break within the expected life of the die. When the presence of lead is suspected, it is customary to add large quantities of pure Kirksite to reduce the percentage of contamination rather than to try to remove the lead.

41. When long runs are to be made that may exceed the expected life of a single die, it is good practice to cast two or more dies at the same time. One die is then used until the distortion becomes excessive and then the second die is mounted on a hammer. The work is brought roughly to shape on the first die and then given a final blow in the second die, which brings the piece to its true size and shape.

42. Kirksite may be obtained in either rolled or cast form, the physical properties and characteristics of each being somewhat different as shown in Tables I and II.

TABLE I

PHYSICAL PROPERTIES OF KIRKSITE AS COMPARED TO SAND-CAST ALUMINUM ALLOY

	Ultimate Tensile Strength p.s.i.	Brinell Hardness	Shrinkage Inch
Kirksite:			
Sand-cast	37,800	100	.14
Rolled	62,000	100	
Aluminum Alloy:			
Sand-cast	19,000 to 29,000	40 to 70	.156

TABLE II

PHYSICAL CHARACTERISTICS OF SAND-CAST KIRKSITE

Comparative strength	60,000 to 75,000 p.s.i.
Melting point	717° F.
Weight per cubic inch	.25 pound
Weight per cubic foot	432 pounds
Elongation in 2 inches	3 per cent
Elongation of rolled Kirksite in 2 inches	30 per cent

43. Dural.—The large amount of scrap dural has promoted the use of this metal for use in cast dies. The dies are light in weight and are tough enough to serve their purpose, having a tensile strength ranging from 33,500 to 37,250 p.s.i., which gives a die somewhat better than one cast of pure zinc but not quite so satisfactory as one made from Kirksite. Scrap dural has a melting point of 1200° F. but considerable dross is formed because of the tendency of this metal to oxidize. To combat this oxidizing, the scrap metal is pressed into briquettes and these are submerged in molten metal left in the bottom of the pot for this purpose.

44. Pouring temperatures range from 1220° F. to 1350° F., but the actual temperature at which the pour takes place should be held between these two extremes. Pouring equipment loses heat rapidly, so that at a pouring temperature too close to the lower limit, 1220° F., the metal may chill as it enters the mold. If for any reason the pouring temperature must be held close to the lower limit, the mold must be skin-dried by means of a torch. Pouring temperatures above 1300° F. usually create bubbling in the mold, causing imperfect castings, and as such a high temperature is not actually necessary it is seldom used. The most satisfactory pouring temperature has been found to be approximately 1280° F., which seems to satisfy all conditions.

45. The shrinkage of dural castings is generally taken as $\frac{5}{32}$ inch for 12 inches and thin checking dies to be used for checking formed parts can be cast and smooth faces obtained with this metal. However, care must be used in making the pour. Enough of the molten metal is poured into the mold to entirely cover the face to a depth of about 1 inch and the pour is then stopped until the face has had time to chill. Additional metal is then poured to a depth of approximately half the depth of the mold and this is allowed to stand until the face has chilled, when the remainder of the mold may be filled. The aluminum will have a tendency to corrode the pot, which must therefore be coated with pipe clay. Under normal operation this coating must be renewed about once a week.

46. Plaster Mock-Up.—Before a mold can be made from which to cast a die it is necessary to have an exact understanding of the characteristics of the metal from which the die is to be made, especially the shrinkage to be expected. With this information at hand, a model or sample piece, known as a mock-up, is made up of wood or plaster. The plaster model is more generally used as it is easier to make and is not subject to distortion by change in atmospheric conditions or by warping. A plaster mock-up costs from one-half to two-thirds as much as a similar mock-up made from wood. If, however, it will be necessary to use the mock-up as a blue block, wood or metal must be used as the bluing will soften the plaster and consequently leaves the surface subject to distortion. The layout is made on a sheet of galvanized material which has been covered with a coating of layout paint. In making the layout, allowance must be made for the shrinkage of the metal and for this purpose a shrink rule is used on which the divisions represent standard divisions, such as 1 inch, plus an allowance for shrink.

47. Templets for a plaster mock-up are cut representing sections through the required part and these sections are welded together to form a skeleton of the part. Clay or plaster is forced into this framework and smoothed off so that the edge of the templets will show, but the surface between them will present a solid body having the required form. After hardening, the surface is carefully smoothed off, for even a scratch mark left in the model will be reflected in the die and cause trouble. The plaster from which the model is made is porous, so that it must be covered with a coating that will fill the pores and present an absolutely smooth and moisture-proof surface. For this purpose a mixture of stearic acid and kerosene, brought to a consistency that can be painted on the model with a brush, is used. The accuracy of the finished part will depend on the accuracy with which the model is made and a correct knowledge of the shrinkage to be allowed for.

48. Zinc and zinc alloys in long thin strips will shrink approximately $\frac{5}{16}$ inch per foot of length, but in three-dimensional bodies—that is, bodies having length, breadth, and thickness—

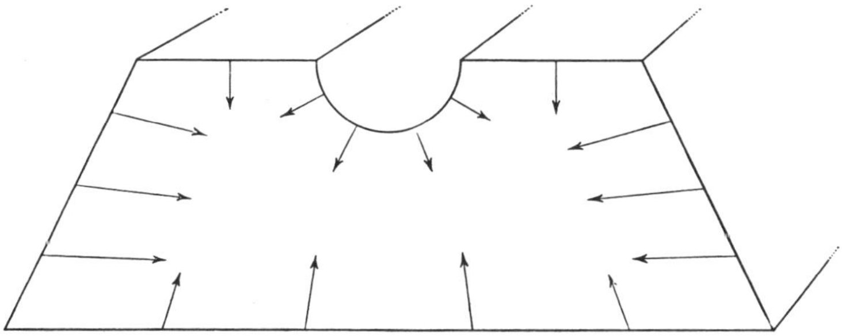

Fig. 8

that are proportional, the shrinkage is always toward the center of the greater mass, in which case the shrink is figured at $\frac{1}{8}$ inch per foot. While a shrink scale is used for general dimensions, it is customary to use a standard scale across openings, as the shrink of the metal will pull away from the opening, giving it the necessary enlargement. The shrink of the metal around an enclosed opening will cause such an opening to close in.

49. The direction of shrink is given diagrammatically in Fig. 8, in which the arrows indicate the direction of shrink for the various surfaces. The amount of shrink at any point will depend on the mass of the metal involved and the shrink for each piece will depend on the shape of that piece. It is only by experience that a thorough knowledge of the exact direction and amount of shrink can be learned.

50. Plaster Patterns.—After the plaster model, or mock-up, has been constructed and the surface protected by a coating of stearic acid and kerosene, it is placed on a level board or platform and boards are used to form a temporary box around it. This box is then filled with plaster, care being taken to force the plaster firmly against the model at all points. When this plaster has set, the box is removed and the plaster cast or pattern of the model lifted away. The same treatment is given to the face of the pattern as was given the model; that is, the surface is carefully smoothed by scraping, sanding, and polishing, after which it is painted with stearic acid and kerosene. The pattern is made with hemp or some similar filler in the plaster to give it addi-

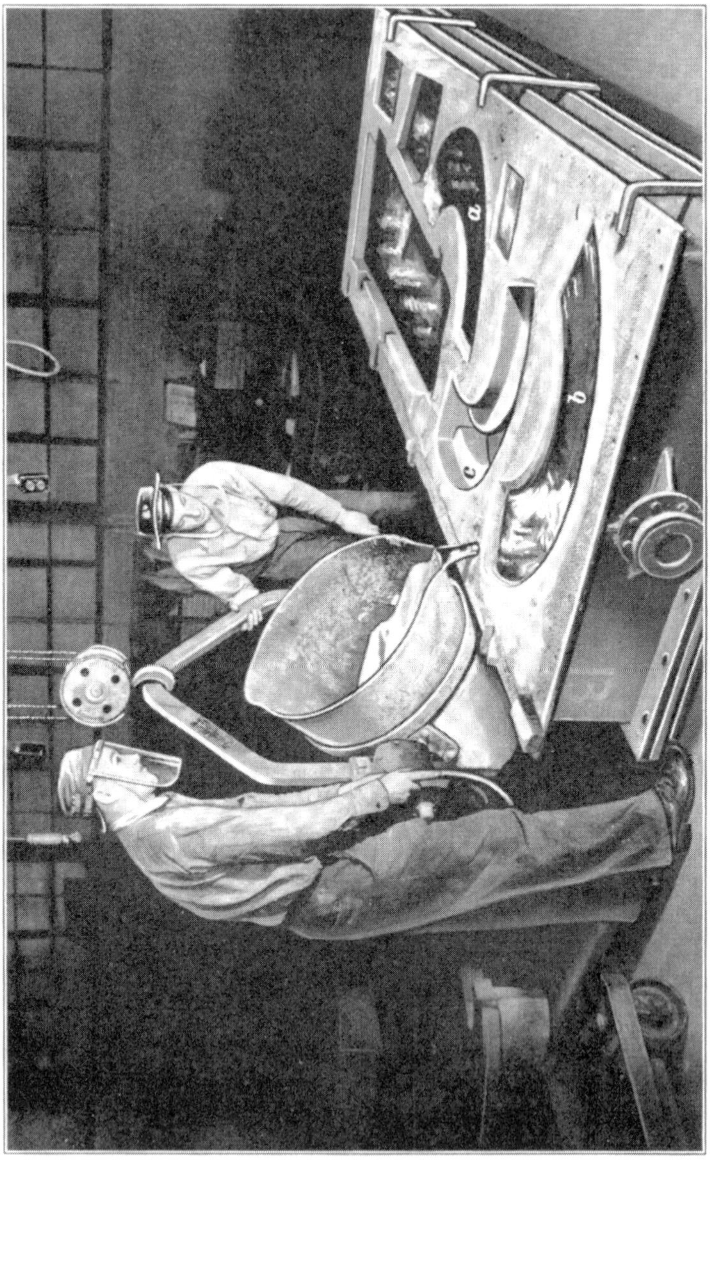

Fig. 9

tional strength. The plaster model is considered positive, since it is exactly the required shape, while the pattern taken from it is considered negative.

51. The negative pattern now has the exact shape of the die by which the finished part is to be formed and can therefore be used as a pattern from which a sand mold of the die can be made. A wall of boards is built around the pattern, forming a box, and this box is filled with molding sand and rammed. Since Kirksite does not throw off gases, there is no need to vent the mold.

52. When the mock-up is too large to be used as a mold, its surface may be laid out into sections and the outline of the sections masked by lines scratched into the plaster. An open box or form somewhat larger than the section to be covered is fitted to the mock-up and is filled with plaster strengthened with hemp or other filler of a similar nature. When the plaster has hardened, this flash, or negative pattern, is removed from the mock-up and is then used as a mold from which a positive pattern can be made. The scribed lines will be reproduced in the negative plaster cast so that the exact size of the section to be reproduced will be indicated.

53. Die Casting.—Provision must be made for handling the die and for this purpose dowels are set in the sand mold so that holes will be formed in the casting. When the casting is handled, steel or iron dowels are inserted in these holes and the hoisting device is attached to them. In some cases it may be necessary to machine the bottom of the die in order to give it a square base, because of the cavitation that takes place when the metal is poured. General practice, however, is to stop pouring when the metal is about an inch below the depth required, and then when the metal has solidified and cavitation has taken place, the last inch is filled in while the metal is still hot enough to make a firm bond. Since the base on which the mold is made is level, the surface of the metal will also be square with the die and smooth enough for all practical purposes.

54. Care should be taken to keep this final layer of metal free of dross by skimming it off as it forms. If the final layer is poured while the rest of the die block is still hot, a perfect bond will result except at the edges, at which a crevice will form. This crevice must be filled with some heat-resisting material such as that used on kettles and ladles and the sides painted with a heat-resisting paint to a height of at least 2 inches above the bottom. This will prevent lead, used to anchor the die to the bed, from entering the crevice and mixing with the Kirksite and will also serve to identify the metal of the die by the color of the paint used.

55. Economy in die making can frequently be obtained by making a number of molds in the same flask, as shown in Fig. 9. Here a number of patterns have been nested together for economy in space. Each mold is gated and is poured separately. In the case illustrated, the pouring is staggered. Mold *a* has been poured nearly full and mold *b* will be filled to the same level, after which mold *c* will be poured. As soon as the metal has solidified and such cavitation as will occur has taken place, additional metal will be poured directly on the surface of the metal to form a level and relatively smooth surface.

56. Die Used as Mold.—The use of a Kirksite die as the mold for a Kirksite or zinc punch is illustrated in Figs. 10 to 14, inclusive. The face of the Kirksite die *a,* Fig. 10, is given a heavy coating of a special material developed for the purpose, which serves not only as a protection to the face of the die but also as insulation between the die and the molten metal, preventing the die from melting or burning from the heat of the molten metal being poured. By giving this coating the thickness of the metal to be formed, the amount of finishing required by the punch can also be reduced. Since all shrinkage is toward the center of the mass, such shrinkage as takes place in the punch will tend to contract it and thereby increase the clearance. This factor must be taken into consideration when the punch is cast from the die.

57. The die *a,* Fig. 10, is placed on a level surface plate and heavy boards, or panels, *b* lined with asbestos are placed around the die to form a flask. The correct thickness for the punch is obtained by means of blocks *c* placed under these boards. Rein-

FIG. 10

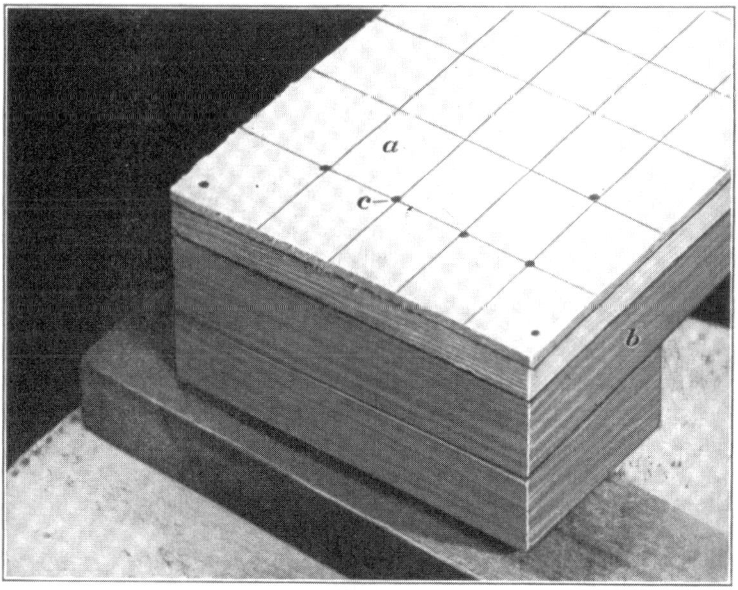

FIG. 11

FIG. 12

forcing blocks *d* stiffen and support the side panels, which are generally made up of two or more narrow pieces. The ends of the panels are staggered so that when in place the exact length of the panels does not have to be the exact length or width of the die. The ends of the panels are then clamped together. Panels varying by 6 inches in length and 2 inches in width will meet all general requirements. End panels are set to a taper of $\frac{1}{2}$ inch per foot of width to provide the proper draft on the casting.

58. The asbestos board *a*, Fig. 11, is built up in two or more layers to a thickness sufficient to protect the wooden side board *b* from the heat of the molten metal and the top sheet is laid off in squares as a means of accurately spacing the nails *c* by which the asbestos is held to the side boards. Proper support must be given the asbestos, as otherwise it may sag or be pulled away from the side board.

59. The completed flask is shown in Fig. 12. Side boards *a* and *c* are placed with one end flush with the end of the die, the

Fig. 13

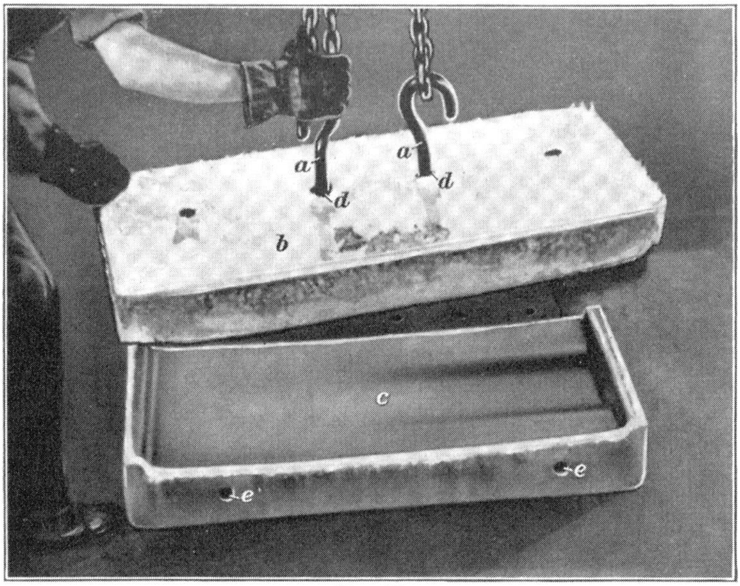

FIG. 14

other end extending beyond the die. An end board *b* is butted against one side board and the second side board *c* butted against the end. The second end board *d*, similarly placed in respect to the side boards, closes the flask. The advantage of this construction is that the boards used need not be cut to fit the die, as any extra length merely acts as overhang. Adjustable clamps *e* are used to hold the boards together, the stiffeners *f* serving as a means by which the boards can be drawn together and held firmly against the die.

60. Some metal is poured into the flask and a spider *a*, Fig. 13, supported at the proper height on the flask by blocks *b*, is put in place. Each bolt *c*, held by the spider, has screwed on to its lower end a threaded socket, or insert, that is imbedded in and held firmly by the metal as it solidifies. The final pouring which brings the die up to the proper level completely buries the sockets. After the metal has cooled the spider is removed, and the bolts are unscrewed from the sockets by means of the hexagon head on the enlarged portion *d* of the screw.

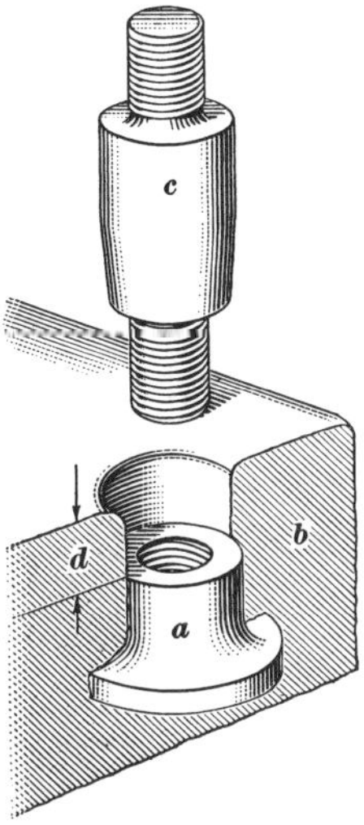

FIG. 15

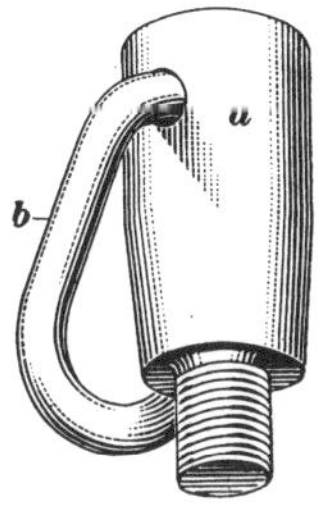

FIG. 16

61. Lifting hooks *a*, Fig. 14, are screwed into the inserts and are used for hoisting the punch *b* out of the die *c*. Four bolts screwed into the inserts *d* are used to hold the punch to the head of the drop hammer when in use. When the die is to be moved, plugs are inserted in the holes *e* and any convenient form of hoist is attached to these plugs. Care must be taken that the type of hoist used is of such a nature that it will not press against the sides of the die and distort it. A variety of hoisting devices have been developed, each of which is suited to a particular type of die.

62. Inserts.—The socket, or insert, *a*, Fig. 15, imbedded in the casting *b* is in the form of a flared-head nut into which is screwed a long stud. A wide collar *c* is formed on the lower end of this stud and when in place the end of this collar is screwed firmly against the insert. The lower portion of the collar is tapered to relieve the friction between the collar and the casting as the stud is unscrewed. The depth *d* to which the insert is imbedded in the casting will depend on the area of the upper face of the punch as well as its weight. For light castings having sufficient surface to permit the use of a number of studs, a thickness of $\frac{1}{2}$ inch is sufficient. In practice a 1-inch thickness is poured so that if necessary this can safely be reduced to $\frac{1}{2}$ inch if the bottom of the casting must be machined smooth. This thickness is increased as the weight of the casting increases.

63. The inserts must be placed so that they will match the holes in the hammer head and consequently the number of inserts that can be used for any given casting is limited. A common hoisting device by which the casting can be handled is a plug *a*, Fig. 16, threaded at its lower end and fitted with a ring or link *b* through the head. The lower end of this plug is given the same taper and diameter as the lower end of the collar *c*, Fig. 15, so that when in place the plug *a*, Fig. 16, it will not only screw solidly against the insert but will also have a tight bearing against the casting. This prevents any motion that might have a tendency to loosen the insert. Enough plugs should be used to prevent excessive strain from occurring on any one of the set that might tend to loosen the insert. Tapered or pointed ends may be screwed onto the ends of the studs to protect the threads that hold the punch to the hammer head as these studs are guided into place.

64. Drop-Hammer Forming.—When a part is formed on a drop-hammer, the action on the blank can be broken down into three separate steps.

1. The sheet metal is forced into the die by the punch so that the metal is actually drawn into the die cavities. Consequently, the metal lying outside of the cavity must be shrunk as it is drawn over the edge of the die.

2. Drawing the metal into the die subjects the outer portion of the blank to compression and, as in the case of a long thin column, this compression causes the formation of wrinkles or waves. The greater the area of the blank lying outside of the die, the greater the tendency to form wrinkles. These wrinkles will be drawn into the die and must be eliminated by shrinking the metal between the punch and the die as the punch bottoms. Wrinkles will be most pronounced at the top of the die where the greatest amount of compression takes place.

3. As the punch bottoms and the wrinkles are worked out, actual shrinking of the metal takes place. All of the metal that lies outside of the die cavity and is drawn into the die occupies a much smaller area, and consequently, unless the die is of such form as to cause some stretching, shrinking the metal causes an increase in thickness. The most pronounced increase in thickness will occur at those points at which the formation of wrinkles is most severe.

65. **Progressive Dies.**—Multistage dies are necessary when surfaces of double curvature having any considerable depth are to be formed. Such dies are modifications of the full-formed die and the pattern for that die must be developed first. Once the full-formed pattern has been developed, such modifications as seem advisable are made, each step being made on the basis of the experience of the pattern maker. The nature of the material being formed must be known and the ability of the metal to form at each step will determine the number of steps into which the forming process must be divided. In the case of excessive forming, heat treating between forming operations may be necessary because of the work hardening resulting from the hammer blows.

66. Three preforming dies may be made for approximately the cost of one finish die, so the saving on the final die will balance this cost. The loss of time consumed in changing the dies between operations will vary from an average of about 19 minutes for a light die to 30 minutes for a heavy die, with a general average of about 27 minutes. Work hardening will occur during forming and between operations so that best results are obtained by annealing the partly formed parts between steps and after the final operation. Also, since annealing is likely to distort the part, it is generally brought to its true shape by a hit after the final annealing has been given the part. The most practical thickness of metal to use for drop-hammer forming is .051 inch. Under a thickness of .025 inch excessive wrinkling takes place and above a thickness of .064 inch the metal work-hardens too rapidly.

67. **Development of Progressive Dies.**—The various steps involved in the development of a progressive die are shown in

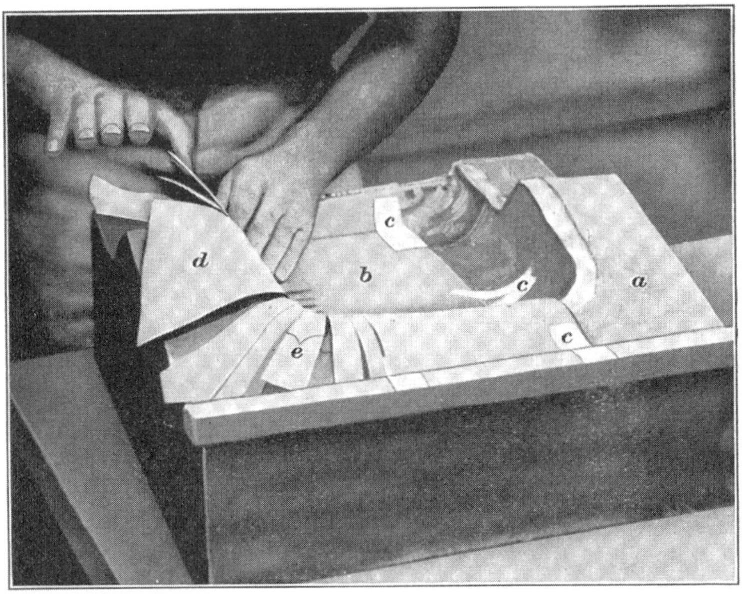

FIG. 17

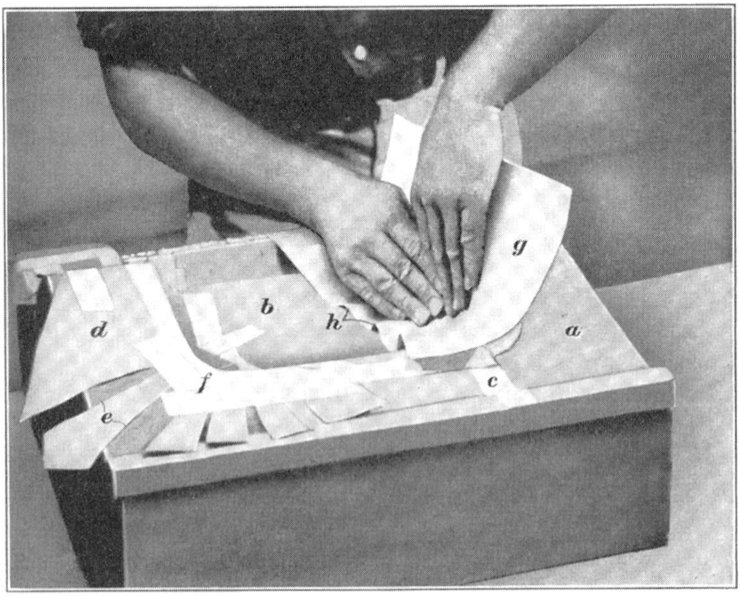

FIG. 18

Figs. 17 to 22, inclusive. The full-formed plaster cast *a,* Fig. 17, is used as the basis for the modified form and a paper pattern is built up in this. To bring the paper to the contour of the proposed die, it is necessary to use a number of separate strips and in some cases to slash these strips so they will fit into the curvature of the plaster pattern. In the case illustrated, a strip *b* just wide enough to fit the straight section is fitted into the pattern and held in place by narrow strips of adhesive paper *c* at the corners. A second strip *d* is pressed into the pattern to get the general shape and slashed at *e* to fit around the corner. As this strip is forced into shape, strips *f,* Fig. 18, of adhesive paper are added to hold the form and to fill in the spaces left open as the strip is bent to the required curvature.

68. A third strip *g,* Fig. 18, is fitted to the opposite side of the plaster cast and slashed to eliminate any wrinkles *h* that may form. This strip is also taped with adhesive paper, so that as a result the entire mold face of the plaster cast is covered by a close-fitting paper pattern. The edge of this paper pattern is trimmed, leaving a sufficient margin *i,* Fig. 19, to form a flange around the pattern. The inside of the paper pattern is given one or more coats of shellac to protect and stiffen the paper and to present a smooth surface. When the pattern maker has determined about how much distortion should be given the paper pattern, he forms a roll of clay *j* over which the paper pattern can be bent.

69. The paper pattern *a,* Fig. 20, is placed over the clay roll *b* and drawn down to the table so that a ridge is formed down the middle. This draws the sides of the pattern in toward the center, eliminating the straight sides of the full-formed pattern, thereby modifying the required forming action on the metal. Clay *c,* Fig. 21, is forced under the paper pattern to support it and additional clay is built up around it to form the necessary body. A bead *d* is formed in the body of the clay that will form a locking bead on the die.

70. Finally, with the aid of a carpenter's square *e,* Fig. 22, a frame of wooden strips *f* is formed squarely around the clay mold

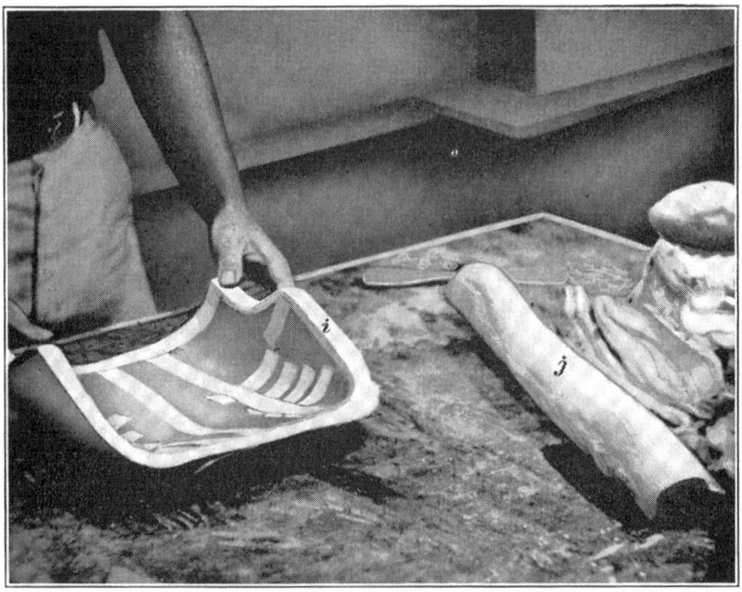

FIG. 19

FIG. 20

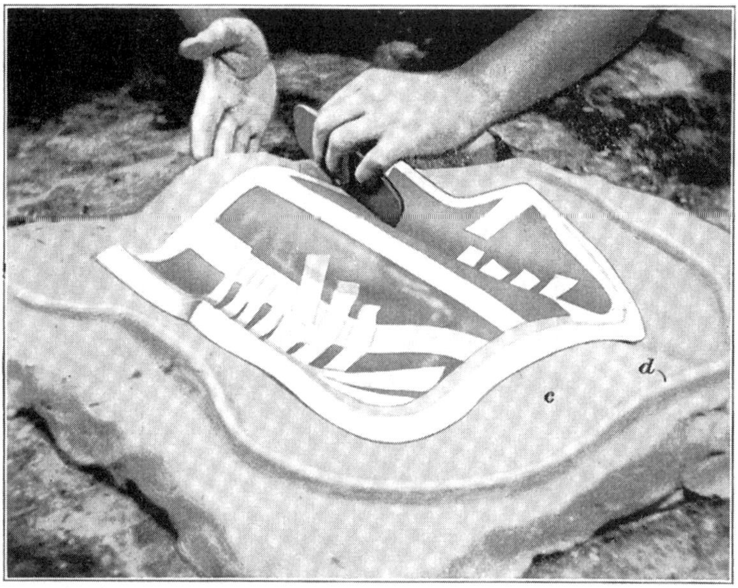

FIG. 21

and any voids filled in with clay so as to form a square die pattern. The paper pattern is then filled with plaster to a sufficient depth to fill the depressions and to build up a block that will be strong enough to withstand the handling it will receive during the casting of the metal dies. The plaster cast, when hardened, is removed and the face is smoothed and treated. It is then ready to serve as a pattern from which a sand mold may be made. The original plaster pattern may be made for either the punch or the die, according to the form required, but in general practice the die is formed.

71. Another use for progressive dies is illustrated in the type of die shown in Fig. 23, which represents diagrammatically a piece being formed to double curvature. Parts that must be formed to double curvature with sharp bends, such as Z or "hat" sections, cannot always be formed satisfactorily in one die because the point of a lead punch will soon become dull. In such cases a double die can be used to advantage and the part reversed so

Fig. 22

Fig. 23

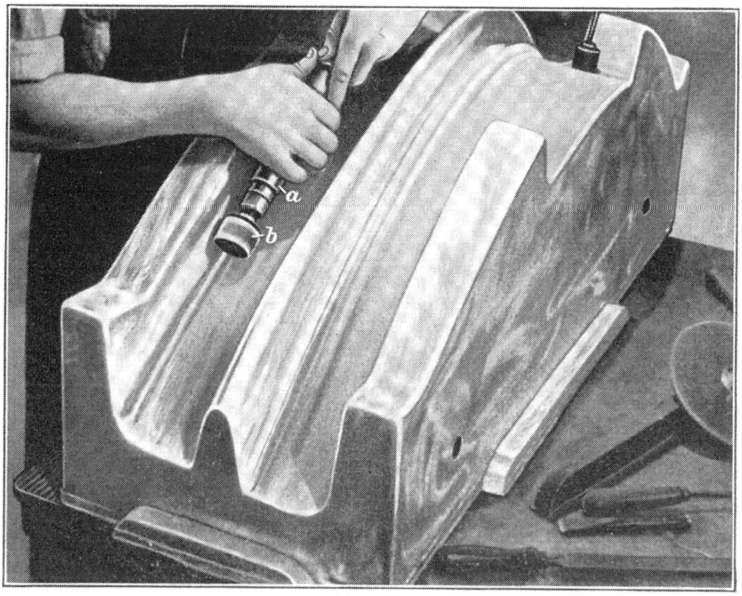

FIG. 24

that it can be hit from both sides. The part *a* is placed in the first die *b* in the die block *c* and is hit, giving it the required double curve. A properly formed punch *d* will give satisfactory results for the first few parts, but if the bend at *e* is sharp, or if the material is heavy, the point forming this bend becomes dull and will no longer force the metal against the die. This trouble is not experienced at the bend *f* in die *g*, since at this point the lead punch continually forms itself. After the first hit has been made the part is turned end for end and placed on the second die. The partly formed bend *e* is now brought to its true shape at *f*.

72. Grinding and Finishing Dies.—After a die or punch has been cast, the working surface must be brought to a mirror-like finish in order that best results may be obtained, and to avoid damage to the aluminum sheet. Also, any excess metal must be removed and all fillets brought to a curvature of the correct radius. At points where metal must be removed, a special tool with an adjustable point may be used to make a scratch of the

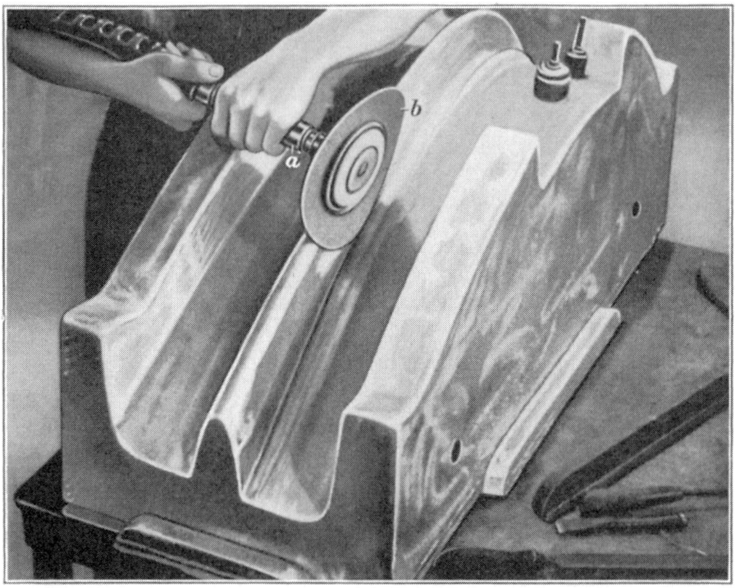

FIG. 25

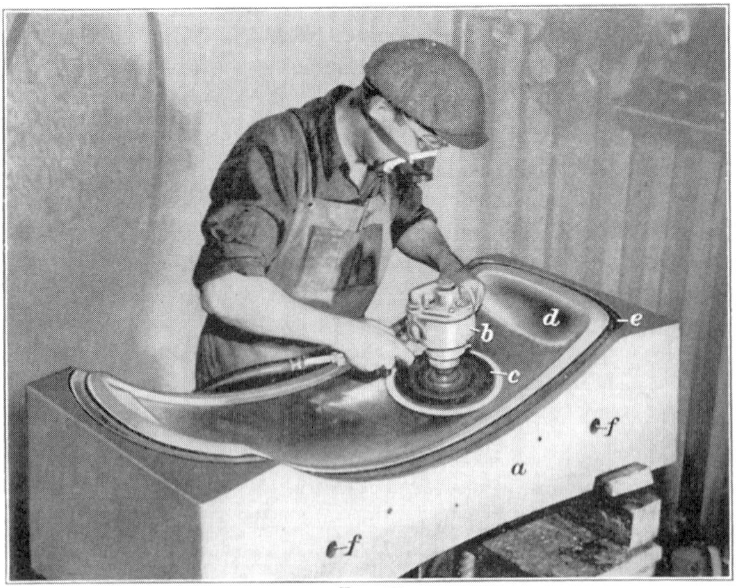

FIG. 26

FIG. 27

required depth and the metal is ground away until this scratch is eliminated. A small portable motor *a*, Fig. 24, to which grinding wheels *b* may be attached, is used for removing excessive metal and for cleaning up fillets to the proper radius; and the same motor *a*, Fig. 25, may be fitted with disks *b* of varying grades of abrasive for finishing and polishing purposes. The final polish is generally obtained by means of a cloth buffing wheel and rouge.

73. An example of a large double-curvature die is shown in Fig. 26. The die *a* has been cast in a sand mold and the punch has been cast from the die. A portable motor *b* and abrasive disk *c* are being used to give the necessary finish to the working face of the die. The finishing of this die is complicated by the indentation *d* at one end and by the groove *e* that matches with the locking bead of the punch. Not only must the face of the die be brought to the proper finish without removing more metal than the proper clearance calls for, but the proper radius of curvature

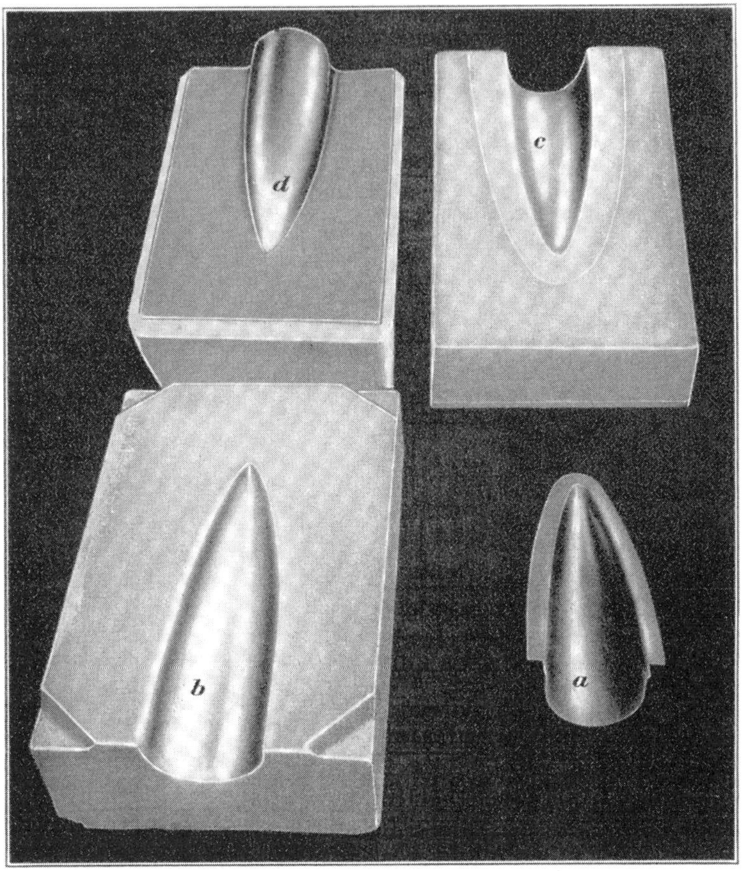

FIG. 28

into the indentation and the groove must be carefully maintained. The holes *f* are cored holes in the casting into which plugs or bolts may be inserted when the die is to be moved by means of a hoist. From 10 to 20 hours is required to finish and polish a set of dies, depending on their size, shape, and contour, when an approximate fit is satisfactory. If, however, the die must be fitted to a blue block, this time will be extended to between 20 and 60 hours.

74. When in use, constant attention must be given both the die and the punch to insure a clean surface at all times. Not only is it necessary to keep the face *a*, Fig. 27, of the punch and the corresponding face of the die clean and well lubricated, but the same care must be given the groove *b* in the punch that mates with the locking bead *c* in the die. The condition of this groove determines the slippage of the blank as forming takes place. If particles of the blank are permitted to adhere to the die, the surface of the next sheet formed will probably be marred or scratched and made unfit for use. Such particles should be detected and removed as soon as they form. These particles can be removed by means of small hand stones or scrapers, as indicated in the illustration; but if the polish of the metal is destroyed, it must be restored by hand by buffer polishing.

75. Example of Cast Die.—The various steps involved in making a cast punch and die are shown in Fig. 28, in which the plaster mold, the punch, the die, and the finished part are assembled. The finished part *a* is a semiconical piece that is to be fitted to the fuselage of a plane. To produce this part on a drop hammer, a plaster mold *b* is first made from engineering drawings or loft sheets, with the necessary allowances for shrinkage provided for. This is a final mold and when used as a pattern for a sand mold from which the die can be cast produces a male mold. The die *c* is cast from Kirksite in the sand mold and when finished is used as the mold into which molten lead is poured to form the punch *d*, which in turn must be finished to allow the necessary clearance required by the thickness of the sheet from which the part *a* is to be made.

NON-METALLIC DIES

CONCRETE AND PLASTER DIES

76. Concrete Dies for Stretch Press.—Many double-curvature contours can be made to advantage on a stretch press. This is especially true when the contour is not great, and when full-length sheets can be used to advantage to eliminate waste. The die may represent the contour of a single part, such as a skin sec-

Fig. 29

tion, or two or more parts, such as right-hand and left-hand fairing pieces, in which case the parts are formed together and then separated. Where long sections are to be formed, waste is avoided by making a die, the base of which will fit closely between the clamps of the stretch press to be used, and the face and curvature of which are such that the sheet is stretched over a curved surface such as a parabola. Such a die is known as a distorted die, since the form to which the sheet is stretched is not the true form. The formed sheet must be removed from the stretch press and straightened out before the double-curvature contour assumes the correct form.

77. The development of a distorted die is based on the lines developed in the Lofting Department. The loft board or sheet on which the contour is shown to full scale by loft lines is sent to the stretch-press department where a wooden check templet is made. This templet, Fig. 29, is made up of an outer frame *a* that gives the exact outline to which the part is to be trimmed, and a number of ribs and cross-pieces *b* spaced closely enough to give the exact contour required. The ribs and cross-pieces are finished to the necessary curvature so that the check templet represents a positive form to which the finished sheet can be fitted. A sheet of metal is hand formed to the exact contour required by this templet and is brought to the shape and contour of the finished piece as it is to be used. The check templet is built up on a solid base *c* which is sufficiently rigid to prevent any distortion of the frame or ribs during the forming of the sheet or during subsequent handling.

78. A sheet of soft steel *a*, Fig. 30, is formed by hand until it fits exactly to the double curvature of the check templet *b* on the base *c*. This work may be done on any type of high-speed hammer in general use for this type of work. The formed sheet is suspended in a framework such that it will assume a curve approximating a parabola, the ends of the sheet being close enough together to form a die that will fit the stretch press on which the work is to be done. Braces are set into the frame so that there will be an open space between the brace and the sheet, and this space is filled in with a mixture of plaster and hemp or

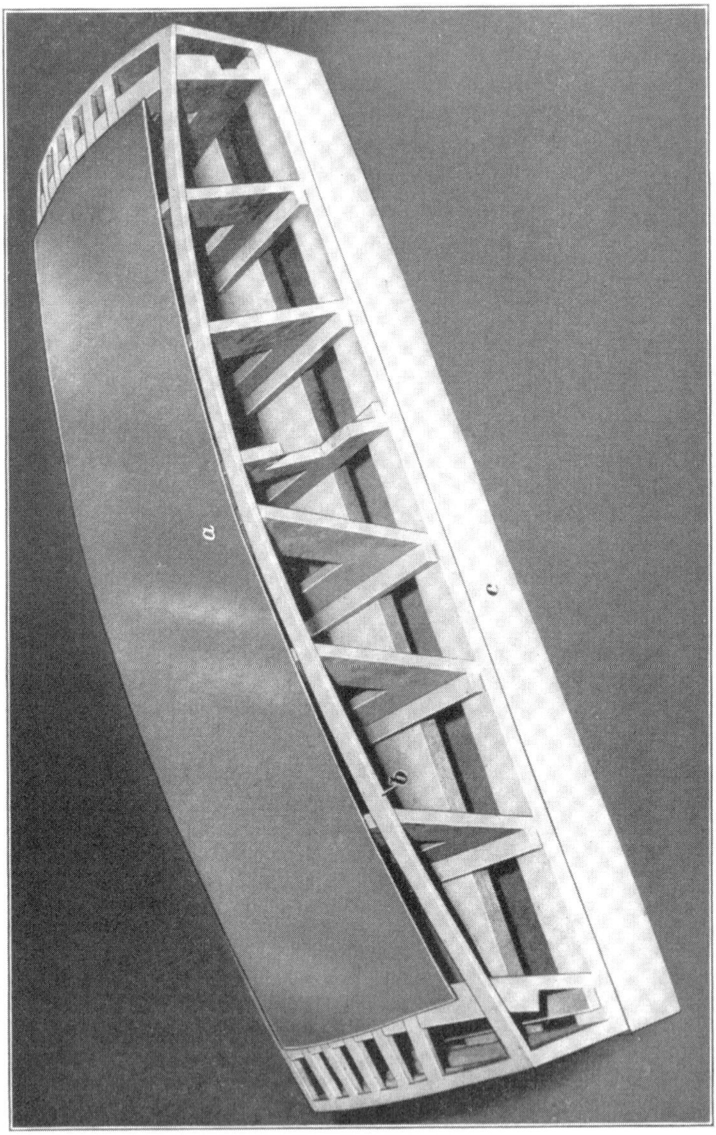

Fig. 30

FIG. 31

some other matter such as excelsior, that will serve to strengthen the plaster. Care must be used, when building up this plaster, not to exert any pressure on the sheet that will distort it from the shape in which it hangs freely in the frame. Bending the sheet to fit the frame reduces the amount of curvature across the sheet until in some cases it practically disappears, or may even become concave.

79. Since the mold is to be pulled away from the sheet, steps must be taken to prevent any adhesion between them. For this purpose the metal sheet is first given a coat of paint and when this paint has dried it is given a coat of stearic acid and kerosene. A layer of plaster about $\frac{1}{2}$ inch thick is immediately troweled onto the sheet. As soon as this layer has begun to harden, a second layer of plaster of about the same thickness is troweled on, and finally a third layer is added. Before the final layer of plaster has completely hardened, a layer of neat cement and water, or grouting, is troweled on and the form is then filled with concrete composed of 1 part of cement, $2\frac{1}{2}$ parts of sand, and $3\frac{1}{2}$ parts of rock, by volume. From 7 to $7\frac{1}{2}$ gallons of water is added for each cubic foot of cement.

80. In order to reduce the weight of the die as much as possible, wooden boxes, or cores may be set in at the base of the die, the size and position of the boxes being determined by the size of the die and the force required in the stretching process. Generally these boxes are made in a U shape to follow the curve of the die. Additional strength is given the die by inserting reinforcing bars around which the concrete is poured. A finish coat of cement can be applied to the top of the concrete as it hardens, to insure a true and smooth base for the die. Since both the plaster *a,* Fig. 31, and the cement *b* have a calcium base, a perfect bond will be formed between the two.

81. The finished die is placed on the stretch press and a trial sheet is stretched. There may be some sag of the sheet between supports because of the weight of the concrete, so this first sheet may not match the check templet exactly and some hand finishing of the die may become necessary. It may be necessary to stretch several trial sheets before the exact contour of the die is obtained, but this loss is more than compensated for by the low cost of the die. On an average, about 15 man hours is required to make a die after the original sheet has been hand-formed. Once the contour of the die has been perfected, a sheet of low-carbon steel *c,* Fig. 31, such as SAE 1010, is stretched over it. This steel serves to protect the face of the die and permits stretching closer to the edge of the die than would otherwise be possible. If too much pressure were exerted directly on the die close to the edges, the plaster would fail and break away. Also, the plaster cannot be smoothed to a point at which some scratching of the aluminum sheet would not result if stretched over a bare die. While these scratches may be invisible to the naked eye, they are sufficient to cause the sheet to weather rapidly.

PLASTICS DIES AND FORMS

82. Introduction.—The term *plastics* is used to cover a number of materials of different types and no single type has all of the desirable properties for tooling. Generally plastics will behave as solids under normal conditions, but when stressed beyond a point known as the yield value they become viscous liquids. Plastics have entered the aircraft industry as punches to be used with drop hammers and hydraulic presses and as forming blocks to be used on hydro-presses. Plastics may also be used for check fixtures, hand-forming blocks for experimental work, tube-bending blocks, and as the base for router blocks. As the nature of plastics material becomes better understood and as plastics for specific purposes are developed their use will become more general in the industry.

83. The development of plastics has been largely in the hands of the manufacturers of synthetic products. Because of their facilities, experience, and general knowledge they are in the best position to carry on the work of developing such special plastics as may be required by the aircraft industry when this development can be carried on in conjunction with the experimental departments in that industry. Plastics can be formed and worked more readily than metal, so that in addition to the saving in critical metals now used, the use of plastics where possible will also make a material saving in the time required for tooling. They have qualities that must be understood, however, to obtain best results.

84. Classification of Plastics.—The classification of plastics is generally based on the reaction of each to heat. The two divisions into which plastics fall on this basis are thermoplastics and thermosetting.

85. The thermoplastic group may be softened and hardened repeatedly by the application of heat and cold without any change in behavior. The basic materials of such plastics are shellac, cellulose acetate, ethyl cellulose, methyl methacrylate, and polystyrene.

86. The thermosetting group is softened during the original heating and hardened by a continuation of the application of heat. Consequently, such plastics, once formed, cannot be altered or reformed by a further application of heat. The basic materials for this group are phenolics and ureas.

87. The basic material, such as cellulose acetate, by which the plastic is known may be the only material used in molding a given product or this material may be mixed with some neutral soluble material, called a plasticizer, that will form a binder for the base material.

88. Fillers.—In general, a filler of some kind is added to the basic material, and these fillers may have a pronounced effect on the physical properties of the plastic. The more general fillers are:

1. Wood flour, generally used with phenolic resins, is finely prepared wood fiber. This is a cheap and satisfactory filler.
2. Cotton flock is sheared from cotton linters and is a very satisfactory filler. It gives greater strength characteristics to the molded plastic than wood flour, but is more expensive.
3. Cloth rags give the molded part a high tensile strength but under compression shears along or parallel to the plane of the filler.

The density of the cloth-rag filled plastic product is somewhat higher than that of either the wood-flour or cotton-flock product, which are about equal.

4. Asbestos and fuller's earth (diatomaceous silica) are also used to some extent.

89. Resins.—There are three classes of resin plastics in general use. They are:

1. "A" stage resins are those that have been brought to the initial stage only, so that the plastic is still soluble and fusible. In this condition the plastic may be used to impregnate other materials. They are thermosetting.
2. "B" stage thermosetting resins have been brought to a point at which they will soften when heated and they will swell when in contact with liquids. They will not entirely fuse or dissolve. These resins are used in molding compounds.
3. "C" stage thermosetting resins have been brought to the final stage and are infusible and insoluble. Such resins are used

in the final molded or cast product. Such products cannot be altered in shape.

90. Phenolic resins are thermosetting and are compounded with an acid catalyst or accelerator. This acid hastens the reaction in the final polymerization or chemical change by which a new compound is formed. The products of the reaction are known as polymers. The addition of the acid to the base resin reduces the time required for work in plastics to about 3 hours, regardless of size.

91. All plastics undergo some shrinkage in their final stage after baking, but this shrinkage can be controlled by the application of continuous pressure during the baking process. The pressure forces the plastic against the walls of the mold all during the jell period. The action of the acid depends on the relative amount added to the basic material. If sufficient acid is added, the plastic will set at room temperature in about 24 hours without baking. The additional acid will cause what is known as oxidation, resulting in turning the plastic dark. Though the finished product will have some additional toughness, very fine surface cracks are likely to result.

92. Application.—Plastics may be used to advantage as the base for masonite form shapes and router block. A flash or wall can be built around the part to make a mold and the plastic formed on the part. After being formed, the plastic is attached or bonded to the part by a layer of the same plastic with additional acid to hasten the reaction. Plastics may be machined by any of the usual tools, such as routers, band saws, and sanders. As plastics are somewhat more abrasive than wood, it is customary to use water as a lubricant during machining operations. If pins or bushings are to be inserted in a plastic form, the necessary holes must be drilled to actual size, since cast phenolic plastics have practically no elasticity.

93. Costs.—One of the advantages of plastics is their relatively low cost, although the low original cost as compared to such metals as Kirksite is somewhat offset by the salvage value of Kirksite dies that are no longer used. Another advantage is comparative lightness of the plastics. A low specific gravity not only makes it easier to handle plastics tools but also helps to reduce costs.

94. Strength.—The strength of a plastics part should be high. The use of plastics as drop-hammer punches, for example, subjects them to severe shock which they must be able to withstand for a reasonable period of useful life. Also, the punch must hold its form and not dull easily at points at which sharp ends are to be formed. The surface of a part must be hard enough to resist abrasion and to retain its original high finish during its useful life.

95. Moisture Absorption.—It is important that a plastic shall be highly resistant to the absorption of moisture; therefore, the plastic should be free of ingredients that would be soluble in or sensitive to water. Any moisture absorbed by a plastics form would tend to cause a change in dimensions or warpage of the finished product.

96. Manufacture.—The acid-setting phenol-formaldehyde thermosetting plastics containing from 25 to 30 per cent of ground walnut-shell flour as a filler finds the widest general application in plastics tools. This material has a wide range of applications as it has excellent casting properties, a high softening point, a short curing period and a high compressive strength. The use of a walnut-shell filler reduces the initial cost to about one-half of that of a corresponding part made from Kirksite. This plastic can be cast in almost any form of mold except sand.

97. The mold should first be sprayed with a hard-finish lacquer that will protect the surface and then given a coating of Barbary wax that will give the surface a high gloss finish that will withstand heavy compression loads. The required volume of plastic is determined from the cavity in the mold. This volume of plastic is prepared and poured into the mold, where it is allowed to stand at atmospheric temperature and pressure for $\frac{1}{2}$ hour to permit the escape of any air bubbles that might be present. The plastic-filled mold is then placed in an oven and held at a temperature of from 100 to 170 °F. and baked. Three hours is generally allowed for baking.

98. Thermoplastic Material.—The most common plastic in the thermoplastic group consists of a phenol ketone known as Plastalloy. It is an amalgamation of phenol and acetone to which is added an acid catalyst, which under the application of heat and pressure copolymerizes; that is, the substances polymerize at the same time to yield a product that is complex, having properties different from either of the polymers alone. The resultant resin remains thermoplastic regardless of the temperature to which it may be subjected. The proper proportion of the ingredients is placed in a cooking tank and thoroughly mixed and melted at a temperature of approximately 205° C., or 400° F., and when melted is allowed to cool to 170° C., or about 340° F. The cooling process will require from 6 to 8 hours, during which time all air bubbles in the mix will rise to the surface.

99. The part to which the plastic is to be cast is backed up with plaster to prevent warping or distortion and is then enclosed in a wall lined with some non-absorbent material, such as tin, to prevent contact between the plaster and the plastic. The high temperature of the plastic would cause any water present to boil, as a result of which air bubbles and holes would be formed in the cast.

100. Pressure is not applied to the cast until a rubbery shell has formed on the surface of the cast as the plastic cools. Pressure is applied by means of clamps and a pressure pad that fits into the flash around the mold, and is exerted while the plastic is still pliable. Hand tightening of the clamps as the plastic cools will exert sufficient pressure to force the plastic to the sides of the mold.

101. Summary of Characteristics.—The advantages of plastics tooling may be listed as follows:

1. Light weight.
2. Easily handled.
3. Quickly and easily produced.

4. Low cost of reproduction of tools, as the original mold may be used repeatedly.
5. Rapid casting to a master part.
6. Plastic will not flow at pressures under 8,000 pounds per square inch.
7. Easily machined with ordinary tools.
8. Replaces critical metals.

102. The disadvantages of plastics tooling may be listed in the following order:
1. Plastics are brittle and must be handled with care.
2. Close tolerances cannot be held in castings of large sizes.
3. Breakage of long, narrow sections will occur unless properly reinforced or backed.
4. May be affected by chemicals.
5. Progressive shrinkage takes place over a period of years.
6. Limited in hydro-press work to approximately .040-inch stainless steel, and .081-inch dural.
7. Pins and bushings cannot be cast directly into such parts as drill plates.

The physical properties of various thermosetting and thermoplastic plastics are given in Table III.

103. Plastics Punch for Die Set.—The construction of a punch and die in which a plastic is used for the punch is shown in Fig. 32. The die *a* may be cast from Kirksite in the usual way, using a plaster pattern and a sand mold. A Kirksite pressure ring *b* is cast and the cavity *c* finished to accept the punch *d*. The punch is made in two parts, a backing plate *e* cast from Kirksite and the punch *d* molded from a thermoplastic material that is somewhat ductile so that the punch will not crack from shock. The punch is held to the backing plate by bolts screwed into inserts imbedded in the plastic. The pressure ring and punch are attached to the ram of a double-action press and the die is fastened to the bed. The blank for the part *f* is placed on the die where it is held by the pressure ring, the corners *g* being formed as the punch forces the blank into the die. The punch is made quite thick so as to take full advantage of the ductility of the plastic and also to allow for a thick and rigid pressure

TABLE III

PHYSICAL PROPERTIES OF PLASTICS

Type	Specific Gravity	Ultimate Tensile Strength P. S. I × 1000	Tensile Modulus of Elasticity × 10^5	Elongation Per Cent in 2 Inches	Ultimate Compressive Strength P. S. I. × 1000	Impact Foot-Pounds per Inch of Notch
Thermosetting Plastics						
Phenol formaldehyde with wood-flour filler	1.3	4 to 11	10 to 15		16.36	.2 to .56
Urea formaldehyde	1.45	5.5 to 7	12 to 15		24	.28 to .32
Phenol furfural	1.3	6 to 11	10.25		28 to 35	.30 to .56
Thermoplastic Plastics						
Methacrylate resin	1.18	4 to 6		1 to 5	10 to 15	.2 to .4
Styrene	1.054	5 to 9	1.7 to 2	2 to 5	11.5 to 13.5	.39 to .50
Vinylidene chloride	1.6	8 to 60	.4 to 2.4	25	15 to 17	2 to 8
Vinyl chloride acetate	1.35	5 to 10	3.5 to 8.5	10.40		.3 to 17
Ethyl cellulose	1.14	6 to 9	1 to 4	10 to 40	10 to 12	.6 to 1.8
Cellulose acetate butyrate	1.14	2.5 to 7.5	2 to 3.5	8 to 8.2	7.5 to 22.3	.8 to 5.5
Cellulose acetate	1.27	3.5 to 10	2 to 3.5	5 to 50	7 to 27	.7 to 4.2
Cellulose nitrate	1.35	5 to 10	2 to 4	10 to 40		2 to 8
Casein	1.35	7.6 to 9	5.1 to 5.7			1.0
Lignin	1.45	9 to 10	15 to 20		25 to 35	.7 to .8

FIG. 32

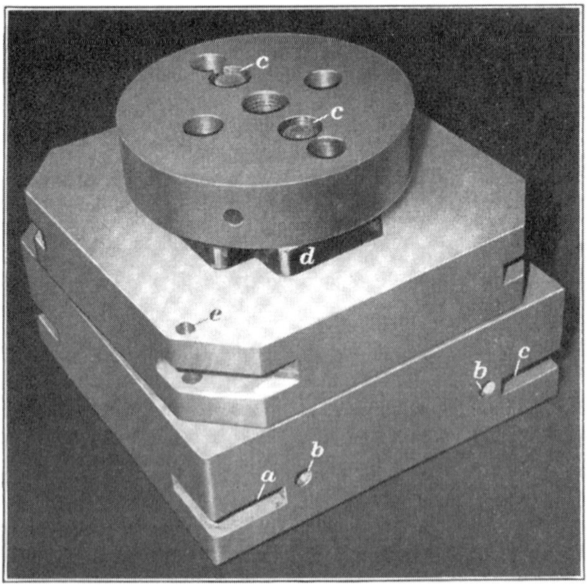

FIG. 33

FIG. 34

plate. Therefore, the punch is supported and restrained by the pressure plate from the backing plate to a line *h* indicating the parting line between the pressure plate and the die.

104. The assembled die and punch are shown in Fig. 33. Clamps are inserted in the slots *a* in the die and are bolted to the press bed in order to hold the die firmly in place. Plugs on the hoisting tackle are inserted in holes *b*, which are cast in the die, when the punch and die are to be moved. Two bolts *c* are used to hold the backing plate to the punch, the holes being countersunk so that the bolt heads will lie below the face of the backing plate. The remaining holes are used in attaching the backing plate to the ram of the press. The punch is long enough to insure ample clearance *d* between the pressure ring and the backing plate when the die set is under full compression force. Pins from the press ram are inserted through holes *e* to raise the pressure plate between operations.

FIG. 35

105. The part *f,* Fig. 32, made from *24SO* having a thickness of .051 inch, is used to form the air duct *a,* Fig. 34, when rolled to shape. A welded joint *b* at each end of the air duct is used to join the ends of the formed piece. Spot welds *c* attach a frame *d* for a diaphragm inside the opening and the quadrant *e* by which the shutter *f* is controlled is riveted to the side. The same part may be made as an all-welded piece *g,* but in this case the air duct must be made up of a number of preformed pieces, necessitating extensive spot welding. The choice between the two methods of making the duct will depend on the machines available, the difficulty in forming the required parts, and the amount of metal involved.

106. Stretch-Press Dies.—Plastics are especially adaptable for use in forming stretch-press dies, since dies of this nature are not subject to shock and plastics have a satisfactory resistance to stress. The mold *a,* Fig. 35, for a die to be used on a stretch press is made on a frame *b.* A sheet of low-carbon steel is

FIG. 36

hand-formed to a check grid, generally made of wooden strips, and is suspended in the frame. Plaster is molded over the steel sheet and gradually built up to the required thickness so that it not only covers the sheet steel but the frame as well. Hemp or some similar material is added to the plaster to give it additional strength and when the plaster has hardened the sheet-steel form is removed. The sides *c* of the mold are made somewhat longer than the required width of the piece to be stretched, so that the

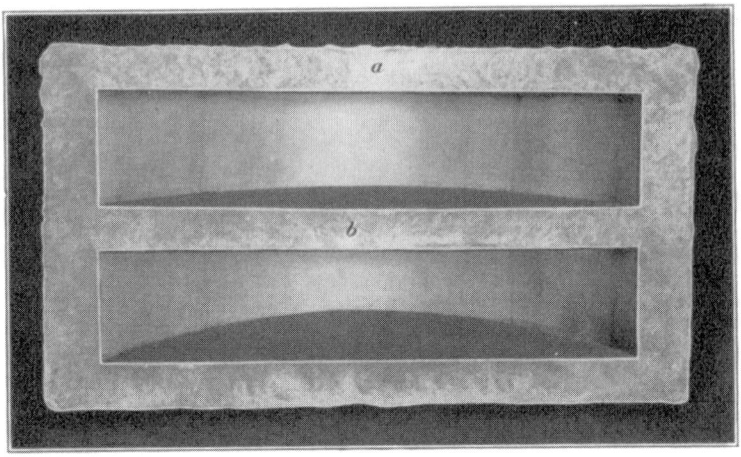

FIG. 37

stress caused by stretching a sheet of aluminum will not come too close to the edge of the die, and the ends *d* of the frame are closed. Any stress that may develop too close to the edge of the mold may cause the edge to crack.

107. In order to save weight and to conserve material, the die is cored into a number of cells, or sections, *e*, Fig. 35, by inserting forms *f* made of wood and sheet metal having the required shape. These forms must be placed so that they will leave ribs *g* that will have sufficient strength to give the necessary support to the die. A thickness of 3 inches is generally allowed for both the shell and the ribs, with a maximum of 20 inches of unsupported shell. There is no definite minimum distance set for the spacing between ribs but a 14-inch spacing is acceptable. The shape of the die may be such that a single rib can be used to advantage with a spacing below this arbitrary minimum.

108. The die shown in Fig. 35 has a ratio of length to width of approximately 3 to 2, so that the cored sections, or cells, are approximately square at the open ends. The side of the cores, being next to the outer shell must, of course, be shaped to the curvature of the shell in order to maintain a constant shell thickness and so avoid unequal shrinking during the curing process.

FIG. 38

The cores are suspended in the frame, leaving a void of the proper width and thickness to be filled with plastic.

109. A phenolic, thermosetting plastic is mixed with about 25 per cent of filler, such as walnut-shell flour, and this mixture is poured into the mold, filling all of the void between the shell and the cores and between the various cores. The mold is placed in an oven at a temperature of 175° F. and allowed to cure for a period of from 2 to 12 hours, depending on the size of the die. After cooling, the cores are removed and the die *a*, Fig. 36, and mold *b* turned over together to the normal position of the die, that is, base down. The mold is removed by attaching a hoist to the frame and lifting it away from the die.

110. Free separation of the die from the mold and cores is provided for by giving them, before the mold and cores are assembled, three coats of a lacquer that serves as a parting material. In spite of the lacquer coating, some plaster *c*, Fig. 36, may adhere to the die and this may be removed by the use of a cloth saturated in a lacquer solvent. The surface of the die will be sufficiently smooth and hard, so that any further finishing will be unnecessary. Wooden blocks are placed in the mold to form recesses *d* in the die that assist in handling it and by which the die can be anchored to the bed of the stretch press.

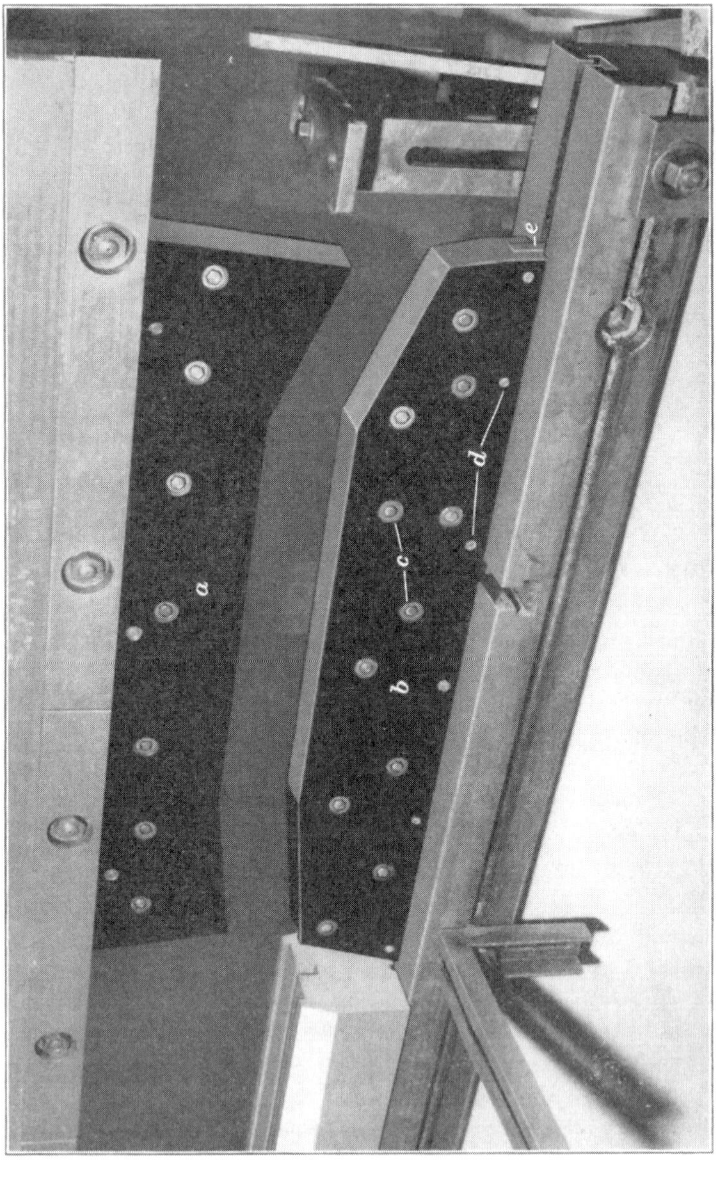

Fig. 39

FIG. 40

111. The closer the ends of the sheet are brought together, the sharper the curve and the smaller the ratio of length to width, but the greater the ratio of length to height. Thus the design of the die will depend to some extent on the machine on which the stretching is to be done as well as the capacity of that machine. A long, narrow die *a*, Fig. 37, may be strengthened by a single rib *b* running lengthwise through the middle of the cavity. In such a case the cavities are formed by two cores having the same

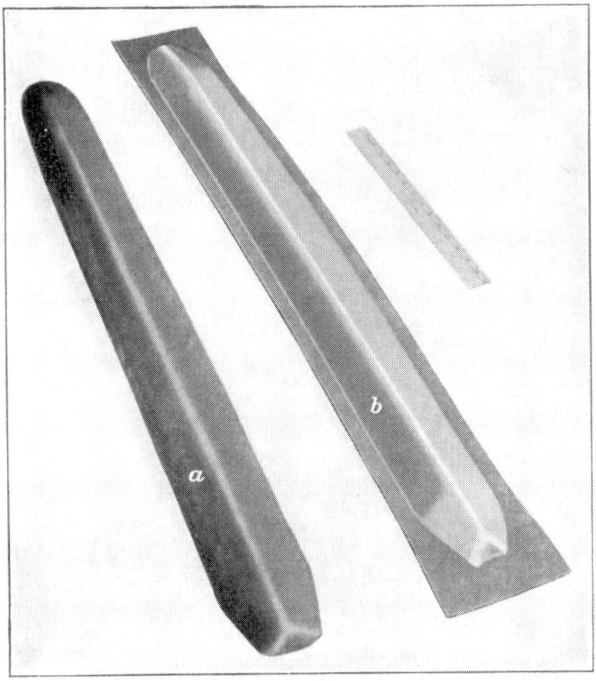

FIG. 41

contour as the face of the die, the proper allowance being made for the required thickness of the shell. Because of unequal shrinkage during the setting and curing of the die, the surface of the ribs and the bottom of the die will be irregular and must be machined to a uniform surface; otherwise the die will crack under stretch pressure.

112. Plastics find their greatest use in the production of small dies, drill jigs, and forming blocks, but they need not necessarily be restricted as to size. A stretch-press die of the form shown in Fig. 38 and weighing 1970 pounds has been successfully produced and used. This same die, if made of one of the lighter metals, would have weighed 16,000 pounds and would also have required considerable time for finishing and polishing. The die shown, which is somewhat smaller, has a length of 7 feet and a width and height of 4 feet.

113. Laminated Plastics.—A type of plastic generally referred to as laminated phenolic plastic is composed of paper laminations impregnated with the plastic material. This forms an extremely hard plastic, but one that is readily machined. It finds its greatest application as forming blocks to be used on either the hydro-press or the press brake. Both the forming block *a*, Fig. 39, and the punch *b* are made of such laminations built up to the required thickness and held together by bolts *c*. Formed into both forming block and punch are T sections *e* held in place by bolts *d* that fit into the T slots of the bed and ram of a press brake. The holes for the bolt-heads and nuts are countersunk so that both sides of the set present flat surfaces.

114. Another type of forming block, used on the hydro-press, is shown in Fig. 40, in which sixteen forming blocks *a* are set on a gang block *b* on the table *c* of the hydro-press. The advantage of using a gang of laminated forming blocks in this way is increased production at a much reduced cost. In the case of the set-up shown in the illustration, the sixteen forming blocks shown were made at a cost slightly less than that of the original metal forming block from which they were patterned. The cost of production is therefore reduced in the ratio of 16 to 1 and production itself increased proportionately. Another important feature lies in the difference in weight, that of the laminated-plastic forming blocks being much less than that of even a light-weight metal forming block. The usual locating pins *d* are used to place the blank in its correct position and the inserts *e* punch out the required holes in the blank as it is formed.

115. A thermoplastic forming block *a*, Fig. 41, can be used to advantage for cases in which the size and shape of the forming block become a matter of serious consideration. The part *b* is so long and the depth of the formed section so deep that if the forming block were made of metal, it would require the combined effort of two or three persons to handle it. When made of plastic, however, the weight is so reduced that it may be handled by a single person.

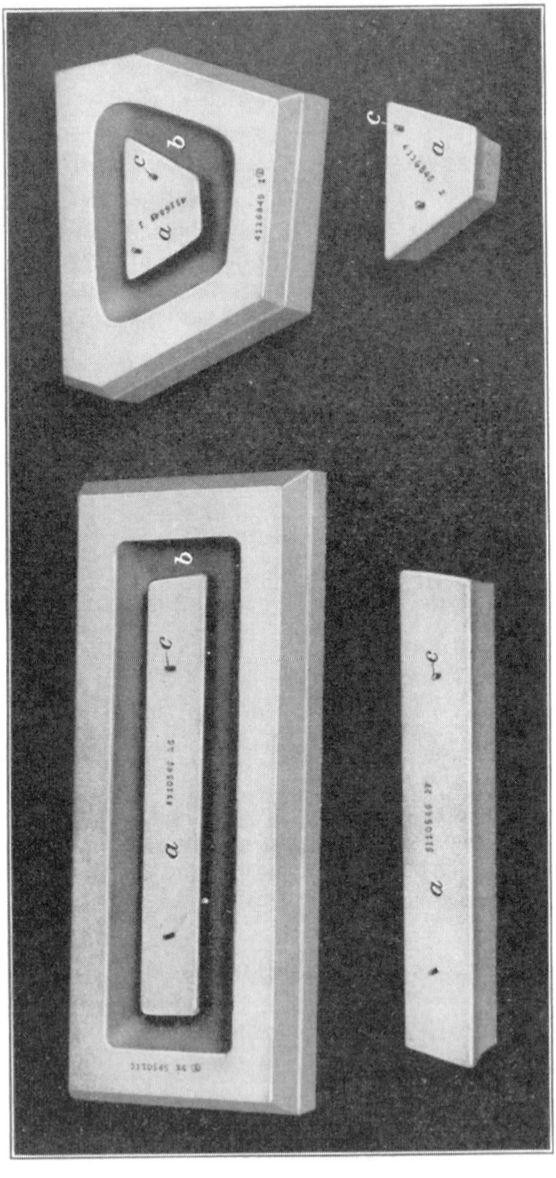

Fig. 42

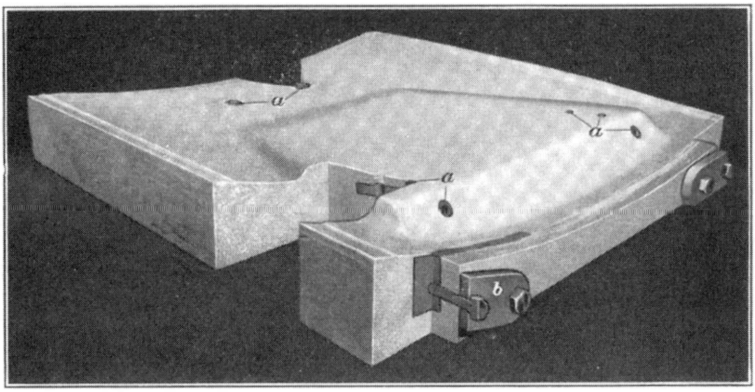

FIG. 43

116. Other types of forming block that can be made to advantage of plastics for use on the hydro-press are shown in Fig. 42. In each case the blank is forced down over the central section *a* of the forming block and either into a cavity *b* of the block or overformed into the undercut portion of the other blocks. Locating pins *c* are used by which the blank is properly placed, and when necessary, pressure plates may be placed on top of the blanks. The customary holes in the pressure plates match the locating pins.

117. Plastics do not have the strength to support screw threads and consequently those points at which metal is to be attached to a plastics form must be reinforced. Also, if a drill jig is to be made from a plastics metal, inserts must be provided that will withstand the abrasive action of the drill. An experimental drill jig, Fig. 43, in which a variety of inserts have been set and sections cut out, shows the methods by which such attachments are held in place. Inserts *a* are bushings set in the block through which the drill passes and attachments *b* bolted to the side of the jig represent metal attachments that must be applied in some cases.

118. The simplest form of insert is a bushing *a*, Fig. 44, machined to a hexagonal form to prevent turning, and with two grooves cut into the shell to form shoulders that prevent the

FIG. 44

bushing from being moved in the block. An oversize hole is drilled at the approximate location of the bushing so that the bushing may be moved about to its correct position. A piece of drill rod is inserted in the bushing and accurately located either from an engineering drawing or by a tooling templet. The void surrounding the bushing is then filled with a metal *b* having a low melting point, such as Cerromatrix. Care must be exercised in selecting this metal, because a definite amount of heat may be generated during the drilling operation and if the melting point of the metal is too low it may soften, allowing the bushing to shift its position.

119. A second and somewhat more involved method of securing drill bushings in place is shown in Fig. 45. A strip of cold-rolled steel *a* is cast in the block *b* at the approximate position of the bushing. An oversize hole *c* is drilled through the block to this plate and a hole drilled and tapped in the plate to receive the end of the bushing *d* that screws into the plate. Care must be exercised not only to locate the bushings properly but also to so place it that it will have the correct angle with respect to the surface, since all holes should be drilled at right angles to the sheet; otherwise, the holes become elongated.

120. A cold-rolled steel insert *a*, Fig. 46, may be used to advantage where metal attachments *b* are to be fastened to the block. The engineering drawings will indicate the location of

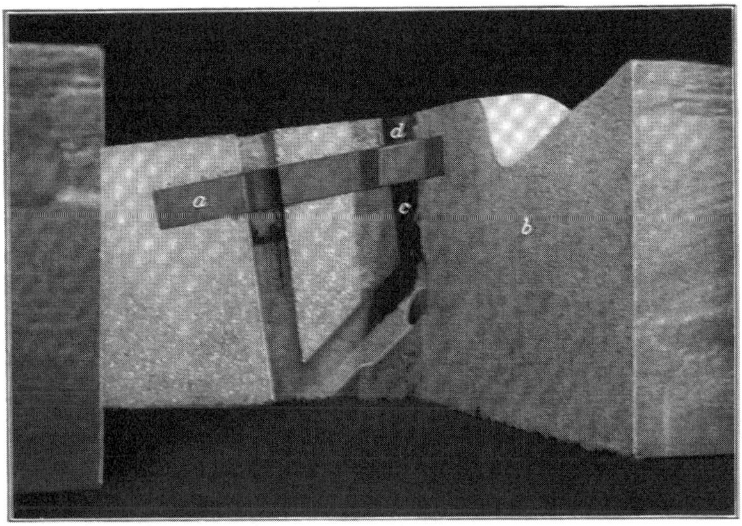

FIG. 45

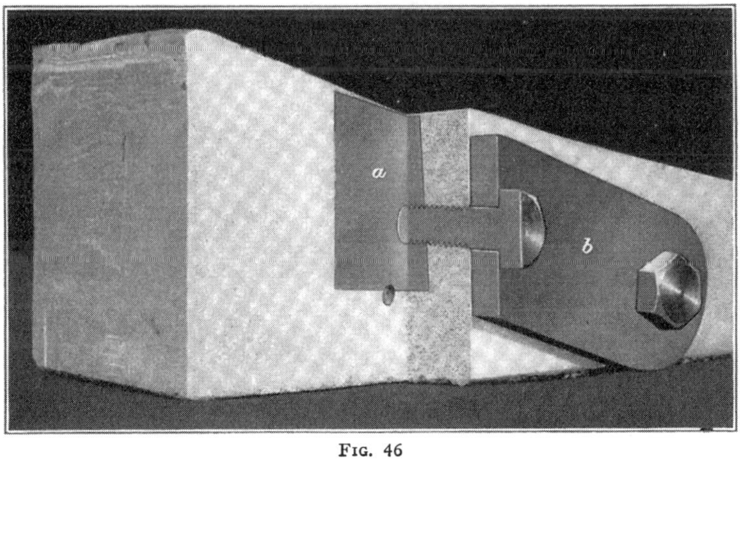

FIG. 46

these inserts. The inserts do not have to be accurately placed provided that full holes can be drilled and tapped through them, with sufficient supporting metal around the threads. The attachment *b* is then accurately located on the block and the necessary holes laid out and drilled. Since the size of drill used will allow for the thread in the insert, it is necessary to tap threads in the plastic, also, and the bolt used must be threaded nearly to the head.